Modern Biotechnology

Modern
BIOTECHNOLOGY

Panacea or new Pandora's box?

Johannes Tramper and Yang Zhu

Wageningen Academic
Publishers

ISBN: 978-90-8686-169-9
e-ISBN: 978-90-8686-725-7
DOI: 10.3920/978-90-8686-725-7

First published, 2011

Cover design & graphics
Identim, Wageningen

Copyright
J. Tramper and Y. Zhu

Wageningen Academic Publishers
The Netherlands, 2011

CONTENTS

7. MEAT FROM THE BIOTECH VAT   113

8. "FRANKENFOOD"   131

PART THREE: HEALTH HAS LIMITS   **151**

9. ANTIBIOTICS   153

PREFACE

There's a long history behind the writing of this book. At its roots are many lectures on modern biotechnology given by one of us (JT) over the last two decades, and still being delivered at universities and schools, in libraries, for service clubs, etc. Many times the question "Why don't you put it all in a book?" was raised. This suggestion solidified in the first half of 2001 when JT took a sabbatical leave at EPFL in Lausanne (Switzerland). A rudimentary draft of the preceding Dutch version of this book was written at that time.

For various reasons this first draft lay pretty well untouched until 2007. In that year the contents of our jobs changed, which allowed us to work structurally on an update and in November 2009 the Dutch version was published. Somebody who greatly facilitated the completion of the book is Tim Jacobs, a young creative graphical designer, who prepared all the figures, cartoons, strips, etc. The regular meetings with him forced us to stay on track and were in fact a great joy. This continued to be the case when he prepared new graphics for this book.

Thanks to grants from Wageningen University and the Wageningen University Fund we were able to widely distribute the Dutch book in the Netherlands. Soon people started to ask: Why didn't you write it in English? Our answer was this: It is born out of the Dutch situation, but we are now working on an international version written in English.

With another grant from Wageningen University, a professional translation was first carried out by Sandra McElroy in a really pleasant collaboration with us. We then produced an updated and international version. Where desirable and appropriate, we replaced typical Dutch cases by international ones and in doing so removed most of the references to Dutch journals, daily newspapers, etc. However, the English text still benefits greatly from pieces of text from Dutch (popular) science writers. We refer to our Dutch book for their credits.

In the present book the number of (complex) links to websites is further increased. To facilitate visiting them they are numbered and the direct links can easily be found on the website of the publisher: www.wageningenacademic.com/modernbiotech.

We gratefully acknowledge the grant from the Netherlands Biotechnology Foundation and the large order by the Netherlands Genomics Initiative; it allowed us to print the book in full colour. We would also like to acknowledge the pleasant and professional collaboration with Wageningen Academic Publishers; having the publisher next door is very handy.

Finally, we should finish with a long list of names of those who, over the years, have contributed in one way or another to the eventual publication of this book. However, the risk of forgetting somebody is so great that we decided simply to issue the following statement: our sincerest thanks to everyone!

The authors

part one

Introduction

In Greek mythology Pandora is the 'giver of all' or the 'all endowed' and the first mortal woman to be sent to earth upon the orders of Zeus. She was given a mysterious box, which she was forbidden to open. Pandora, however, not only possessed the charm and beauty of a goddess, a gift from Aphrodite, she was also very curious, a characteristic given her by the god Hermes. Once on earth, therefore, she was unable to resist taking a look inside the box. It was filled with gifts and calamities, all of which, to her dismay, escaped and spread throughout humanity, with all the disastrous consequences thereof. Only the spirit of hope was left at the bottom. Figuratively speaking, Pandora's box is a source of much suffering. Is modern biotechnology a Pandora's box, as anti-biotechnology movements would have us believe or is it a panacea to cure many of the world's ills? Therein lies the pivotal question in this book. Our final conclusion is that biotechnology can be the source of much good if it is handled wisely; in other words, we should lift the lid of this new Pandora's box carefully and with discretion.

PANDORA'S BOX...

MODERN BIOTECHNOLOGY:
A BLESSING OR A CURSE?

Developments in the area of modern biotechnology can no longer be stopped. Take, for example, the amazing pace at which our knowledge and understanding of the genetic material of humans (Textbox 1.1) is moving and the possibilities that this opens up for health care and forensic science. It's vital to put this into practice in a sensible and controlled manner. Winning the trust of the public must be the first step. But reliable information and continuous communication are crucial if that is to happen. In this book we aim to go some way towards achieving this. The main focus is on the more controversial topics, such as gene therapy versus gene doping, or therapeutic versus reproductive cloning. The most famous example of cloning is Dolly the sheep, born in 1996 and the first cloned mammal. In this chapter we aim to make just a passing acquaintance with modern biotechnology for those who are unfamiliar with this fast-evolving area of expertise. We have tried to write the various chapters so that they can stand alone and be read separately. The textboxes contain more detailed information, basic knowledge, or typical examples, but are not needed for understanding the main body of the text.

THE BIRTH OF DOLLY WAS NOTHING SHORT OF A MIRACLE...
BUT VERY SHORT ON ROMANCE!

1.1. WHAT IS (MODERN) BIOTECHNOLOGY?

There are many definitions of biotechnology. Basically they all come down to the same thing: biotechnology is the integration of life sciences with engineering. The production of semi-synthetic antibiotics like amoxicillin by using moulds and enzymes is an excellent example of this. People talk about modern biotechnology when recombinant DNA technology is involved, also called gene technology or genetic modification (see below and also Textbox 2.1 in the next chapter). Opponents of modern biotechnology consistently use the term genetic manipulation because it provokes a negative association. And yet gene technology doesn't attract universal criticism. Quite the opposite. There are many examples of great new products of modern biotechnology and there are many more in the pipeline, especially in the medical domain. The development of genetically modified moulds has, for instance, made the production of antibiotics far more efficient in the last few decades. Since mid-1990, however, there have been heated discussions among supporters and opponents of modern biotechnology, especially in Europe. The thorny issues in these debates have been estimating the risks in terms of health and the environment. It seems that, for the time being at least, it won't be easy to close the gap that has opened up over the years between the various points of view. The European debate is a good reflection of what is happening around the world. For example, Prince Charles of Great Britain is an out-and-out opponent while Jimmy Carter, ex-president of the US, and winner of the 2002 Nobel Peace Prize as well as founder of the non-profit Carter Center[1], is a major supporter. The Vatican gives a cautious nod to biotechnology, on the understanding that there should continue to be a ban on cloning humans and 'tinkering' with human DNA. It is also clear that even renowned scientists cannot seem to agree with each other and nobody can guarantee absolute safety.

The burning question is whether biotechnology and its products are more dangerous than more conventional equivalents and whether they fit into the picture we as a society have of the future. For us to be able to establish this, we must keep up the societal debate and continue to research and develop modern biotechnology.

1.2. BIOETHICS

At the beginning of 2002, Francis Fukuyama's book *Our Posthuman Future: Consequences of the Biotechnology Revolution* (Fukuyama, 2002) was discussed in many book review sections of newspapers, magazines and journals, e.g. Abrams (2003) and Spier (2002). Francis Fukuyama is certainly not a run-of-the-mill thinker. He is a renowned political philosopher, and at the beginning of this century – as a member of the presidential advisory council on bioethics – he directly advised George Bush on matters such as cloning, the use of embryonic stem cells and genetic selection and modification. His opinions therefore partly determined how the Bush government

[1] www.cartercenter.org

TEXTBOX 1.1.

Structure and function of genetic material.

Genetic information is stored in DNA molecules (deoxyribonucleic acid). A DNA molecule is a long strand of nucleotides which are linked to each other by phosphate groups (the black balls in Figure 1.1). A nucleotide consists of a deoxyribose molecule (the sugar ribose in which an –OH is replaced by –H) to which a nitrogen base is attached. DNA contains four different nitrogen bases: adenine (A), cytosine (C), guanine (G) and thymine (T). Genetic characteristics are established as genes in the DNA molecules. A gene is a piece of a DNA molecule that codes for a specific protein. In other words, if a cell contains DNA with a specific gene, this cell can theoretically make (express) the protein encoded by this gene. Proteins regulate all processes in the living cell and as such are the building blocks of life. They themselves are made up of 20 smaller units, called amino acids.

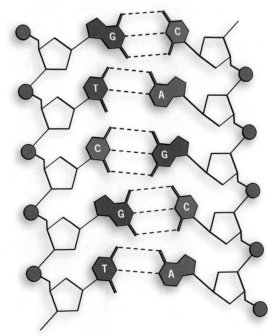

Figure 1.1. The two-dimensional structure of DNA.

The sequence of the nucleotides is the code used to lay down genetic information. This code is always laid down in sets of three building blocks, the so-called triplet code (Figure 1.2). One triplet is called a codon and represents one amino acid in the amino acid chain from which proteins are then produced. Most amino acids have several codons and there are also stop codons. This doesn't mean that the whole DNA molecule is used from start to finish to code genetic information. On the contrary, these codes are distributed in small packages on the DNA, the genes. Other pieces of DNA are located between the genes, and these function like dimmer switches, regulating the action of genes or groups of genes. In addition, between the genes there are much bigger pieces of DNA whose function we do not yet know, and which have long been thought to have no function at all. This view, however, is increasingly being challenged. These pieces are still frequently, and unjustly, called junk DNA.

The protein synthesis is carried out by ribonucleic acid (RNA). This consists of a single stranded chain, similar in structure to that of DNA, the difference being that deoxyribose and thymine are replaced by ribose and uracil (U), respectively. RNA is made in the cell nucleus on the DNA (transcription) and occurs in three forms. Messenger-RNA (mRNA) takes the information necessary for the protein synthesis from the DNA in the nucleus to the protein factories of the cell, the so-called ribosomes. Transfer-RNA (tRNA) is the form which transports amino acids to the ribosomes and sequences them along the mRNA. For each amino acid there is a separate tRNA molecule with a specific triplet of unpaired bases, the anticodon. This pairs with the corresponding codon in the mRNA molecule in the ribosome, resulting in the coupling of amino acids to each other to form proteins according to the base sequence in the mRNA (translation). In most organisms ribosomes roughly consist of equal parts of protein and ribosomal RNA, the third form of RNA.

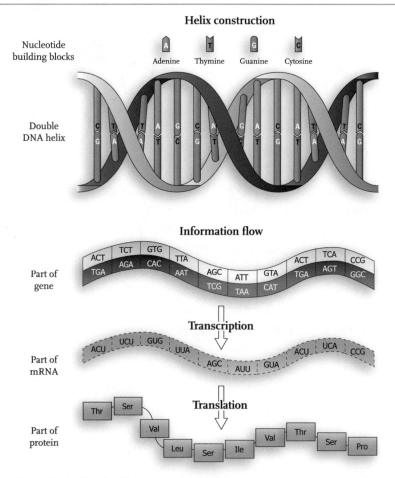

Figure 1.2. The protein synthesis (Thr, Ser, Val, etc. are the separate amino acids in the protein chains).

dealt with modern biotechnology. One example of his influence was the ban imposed by the Bush government on using state funding for embryonic stem cell research (see Chapter 14). According to Fukuyama, state power should be used to lay down the rules for biotechnology, and this should not be left to science or business, nor individual freedom of choice. The consequences of an unlimited application would, in his view, be too drastic and dangerous.

A critical analysis of Fukuyama's bioethics was published by Jordaan (2009). Jordaan identifies four distinct weaknesses in Fukuyama's main argument, i.e. human nature, which is defined as species-typical genetic characteristics, is the ultimate basis for human values, specifically for our species' special value – our human dignity. Fukuyama infers from this first premise that should any aspect of human nature be changed by new reproductive technologies, it would endanger not only human values, but also the very basis of human dignity; therefore justifying the limitation of such new reproductive technologies to therapeutic uses. The four weaknesses Jordaan identifies are: (1) Fukuyama's definition of human nature is vague and not based on reality; (2) the relationship he posits between human nature on the one hand and values on the other is weak, or dependent on other, non-related values; (3) even accepting his first premise, it does not follow that any change in human nature will necessarily undermine human dignity; and (4) even accepting his second premise – any change in human nature will necessarily undermine human dignity – it still does not follow that new reproductive technologies

should be limited to therapeutic purposes. Jordaan therefore submits that Fukuyama's main argument must be rejected. Subsequently he analyses the supporting arguments in *Our Posthuman Future* – relating to reproductive freedom and human rights, social justice, and psychology – and concludes that none of these arguments can support Fukuyama's contention that new reproductive technologies should be limited to therapeutic purposes. Jordaan ends his thirteen-page analysis with the paragraph *Antipromethean Heresy:*

"*Our Posthuman Future* is a good dose of feel-good drugs in a philosophical sugar-coating for a bioconservative audience. To a more open-minded, ethically informed audience, *Our Posthuman Future* is a macabre effort to resurrect the discredited naturalistic fallacy back into mainstream philosophical discourse after a well-deserved death more than a century ago ... *Our Posthuman Future* cannot add anything to the global bioethics debate – it can only pollute these already troubled waters with arguments that have the intellectual accountability of a tabloid feature. Dworkin (2002) describes the ethical dimension of the biotechnology revolution as 'playing with fire', but states that playing with fire is 'what we mortals have done since Prometheus'. But then, not all philosophers can have promethean courage to face and explore a radically new value paradigm. Fukuyama clearly prefers humankind to live without this metaphorical fire. The promethean metaphor has been a defining paradigm in classical times, as well as in modernity – it was the cultural catalyst for creating the free and technologically advanced contemporary societies of the West … Beware the day when we betray our promethean heritage. Beware the antipromethean heresy of Fukuyama."

Although we are not professional philosophers, we subscribe to the view of Jordaan. We believe that only open debates by a well-informed public, hand in hand with ongoing education, scientific research and technology development, can create a viable future for humankind.

In a leading Dutch newspaper (*NRC*), two years before the publication of Fukuyama's book, the following headline appeared above an article on the opinion page:

Biotechnology is not harmful

The article was written by Cynthia Schneider, who at that time was the US ambassador to the Netherlands (she left in mid-2001). Schneider wrote this article to publicise an international conference she had organised to take place a few days later in The Hague. A symposium attended by a great many of the big shots in the area of modern biotechnology, among them J. Craig Venter, who came to champion her proposition. As a result of a false bomb threat issued by an opponent of modern biotechnology, the symposium was temporarily suspended. We are not opponents of biotechnology. On the contrary, it has been our

Part 1: Introduction

professional field for decades. But neither are we as decided as Mrs. Schneider. Using the same words, we would like to convert her statement into a question:

Isn't biotechnology harmful?

And that is exactly what we will be discussing in this book so as to be able to arrive at the final conclusion: Biotechnology doesn't have to be harmful! If used sensibly, it can be a blessing rather than a curse for mankind. Or, in the words of Richard Preston, author of the bestseller, *The Cobra Event* (1997):

I don't want 'The Cobra Event' to be seen as anti-biotechnology or anti-science, since it isn't. In the introduction I compare genetic engineering to metallurgy – it can be used to make plowshares or swords. The difference is human intent.

1.3. BIOTERRORISM

Preston's book, sadly enough, is about a terrorist attack on New York City. Not with aeroplanes, but with a genetically modified virus. The plot rings so true, that former president Bill Clinton asked FBI experts to look into how realistic such an attack was. Opinion polls in the US, carried out online in 2000, revealed that 94% of respondents were worried that their country was vulnerable to attacks by bioterrorists; 64% even thought there was a serious risk of the attack taking place in the first decade of the 21st century. A

convincing response! After the terrible attacks on the Twin Towers in 2001, the fear of biological attacks in particular has continued to increase and many American citizens have even gone so far as to buy gas masks. More and more countries are therefore trying to prepare for an attack with biological weapons, whereby ironically biotechnology itself will probably also be used to ensure defence. Practically speaking, the research in question often amounts to the same thing as developing these weapons. In mid-2007 five American laboratories conducting this kind of research were closed down, because staff there were infected with offensive pathogens. There is also a suspicion that a lot of defensive work against bio-weapons is offensive in nature. For more information on this topic, go to the website set up by the Sunshine Project, an international non-profit organisation that works to bring to light facts about biological weapons[2].

1.4. RECOMBINANT DNA TECHNOLOGY

The year 1973 was a special year. It was the year of the energy crisis resulting in car-free Sundays. And of course, it was also the year of the Watergate affair which led to the resignation of the American president Richard Nixon. But this year is also regarded as the birth of modern biotechnology. It was in 1973 that Stanley Cohen and Herbert Boyer[3], of Stanford University and the University of California in Berkeley, carried out the first successful recombinant DNA experiments with the

[2] www.sunshine-project.org
[3] web.mit.edu/invent/iow/boyercohen.html

TEXTBOX 1.2.
A "triple lock" on the door!

Above it was mentioned that in 1973 Berg, Boyer and Cohen were the first to "recombine" an organism. At the time a discussion was already brewing about the risks of this new technique. At the request of the National Academy of Sciences, Berg sent a letter to the journal Science, in which he called for a one-year moratorium on further recombinant DNA experiments (Berg et al., 1974)[4]. Together with about 150 other scientific experts, including one of the two scientists who discovered the double helix structure of DNA (Figure 1.2), James Watson, Berg formed a committee to discuss the potential risks of this technology. These discussions were made public at a conference in Asilomar, California, in February 1975, where guidelines for safe experimentation were laid down. In the first instance, these guidelines only concerned microorganisms (bacteria, yeasts and moulds). Special instructions for plants and animals only came later. The recombinant DNA laboratories are literally equipped with locks and only specially trained researchers are allowed to enter. In these laboratories, a vacuum, amongst other things, is supposed to prevent microorganisms from 'escaping'. 'Crippled' microorganisms are used, so that in the event that they do escape, they will not be able to survive 'on the outside'. These measures would appear to have been overcautious and have since been relaxed.

---

bacteria *Escherichia coli (E. coli)*. With genetic "copy and pasting" they made this *E. coli*, a bacterium that lives in our intestines, resistant to two antibiotics, namely tetracycline and kanamycin. In that same period their colleague and later Nobel Prize winner Paul Berg (also at Stanford University) modified the genetic material of the same microbial strain with a piece of DNA from a cancer-inducing virus. These scientists foresaw the enormous consequences of this new technology and called for a voluntary, temporary moratorium on further research. Once guidelines for safe experimentation had been established during the Asilomar conference (Textbox 1.2) in 1975, research in this area took a great leap forward.

[4] www.pnas.org/cgi/reprint/71/7/2593

Since then, the term modern biotechnology has been used whenever recombinant DNA technology is applied (see Textbox 2.1 in Chapter 2 for more information on this technology).

The first commercial applications of this technology followed less than a decade later, in 1982. The Dutch company Intervet was the first on the market with a vaccine against swine diarrhoea. Hot on their tails was the American company Eli Lilly with human insulin for diabetics, made from genetically modified bacteria (*E. coli*). A piece of human DNA - the bit that ensures we can make insulin in our body - was added to the DNA of these bacteria, so that these microorganisms could make human insulin for us. As a result, unlimited quantities of pure human insulin, so to speak, have been made available at a reasonable price. Furthermore, this insulin causes fewer side effects than the old product, i.e. modified swine insulin.

1.5. BIOTECHNOLOGY DEBATE

When recombinant insulin appeared on the market, there immediately arose a heated debate between supporters and opponents of modern biotechnology. A German company had developed a similar commercial process at the same time, but only got manufacturing permission from the German government a good ten years later due to pressure from the Green Party. The Green Party and other environmental groups forced the introduction of very restrictive German legislation concerning modern biotechnology, because of the fear of irreparable damage to the environment and health. However, German diabetics protested that they should also be able to access this new medicine directly. This led to a hypocritical situation where the product couldn't be manufactured in their own country, while at the same time it was being imported and sold on the market. The debate about this technology has continued ever since, and has become even more heated since mid-1990. It would be no exaggeration to say that we have ended up in a situation of trench warfare. In the various chapters of this book we will be reviewing in particular the more controversial topics like recombinant products as food additives, genetically modified plants and animals, and cloning. These are discussed in the context of our daily food and drink (Part II) and our health (Part III).

There's no shortage of coverage of gene technology in the media. While lecturing around and about the country, however, it has become clear to us that we scientists, but also those in business, the public sector and the media, have not yet succeeded in conveying sufficient knowledge in the area of modern biotechnology to the man in the street. Which is why fear of this technology has sometimes been blown out of all proportion. We are definitely not in favour of indiscriminately implementing everything humanly possible with the help of modern biotechnology. We do, however, believe that wise use of gene technology can lead the way in developments that will create new and better products. First and foremost it is essential to establish standards and norms for implementing gene technology, such that the man in the street starts to believe that these developments can proceed safely with no risk to our health and the environment. With this book we hope to objectively inform the wider public about modern biotechnology in order to reach the point where the discussion can turn to the real issues. Not those that are chiefly dictated by irrational fear and end up in a "yes it is/no it isn't" discussion. But rather, what do we as a society consider to be acceptable risks and which objectives do we classify as sufficiently important to justify the use of modern biotechnology.

1.6. SOURCES

Abrams, F. R. (2003). *JAMA*, *289*(4), 488-490.

Berg, P., Baltimor.D, Boyer, H. W., Cohen, S. N., Davis, R. W., Hogness, D. S., *et al.* (1974). Potential biohazards of recombinant DNA-molecules. *Science*, *185*(4148), 303-303.

Dworkin, R. (2002). Sovereign virtue: the theory and practice of equality (First paperback edition ed.). Cambridge, MA, Harvard University Press.

Fukuyama, F. (2002). Our Posthuman Future: Consequences of the Biotechnology Revolution. New York, Farrar, Straus & Giroux.

Jordaan, D. W. (2009). Antipromethean Fallacies: A Critique of Fukuyama's Bioethics. *Biotechnology Law Report*, *28*(5), 577-590.

Preston, R. (1997). The Cobra Event. Toronto, Canada: The Random House Publishing Group.

Spier, R. E. (2002). Toward a New Human Species? *Science*, *296*(5574), 1807-1808.

"It is one thing to have a safe product; it is another to command confidence in the market place"

Stephen Dorrell, former UK Health minister (1995-1997, Conservative party)

Advances in the field of modern biotechnology seem unstoppable now. This view is also expressed in Ernst & Young's annual reports on the biotechnology sector. According to their *Beyond borders: the global biotechnology report 2007*[5], biotechnology even experienced a historical leap forward in 2006 with global growth of more than 10 percent. The growth in global areal with transgenic (genetically modified) crops – in the EU the most controversial issue – has been steady right from the introduction in 1996 (Figure 2.1a) and in 2009 this areal has again grown by an ample 7%, totalling 134 million hectares (ISAAA[6]). The growth is especially strong in Brazil, South Africa and India. Brazil has even overtaken Argentina and is now, after the USA, the largest producer of transgenic crops. This figure also shows the hesitance with respect to transgenic crops that exists in Europe; the area occupied by these crops there is marginal (less than 0.1% of the global total). In 2007 this area even halved, which was largely due to the almost complete disappearance of these crops in Romania (Figure 2.1b). Nevertheless, at the time of writing this, in the summer of 2010, it seems that even in Europe the tide is also turning in favour of these crops.

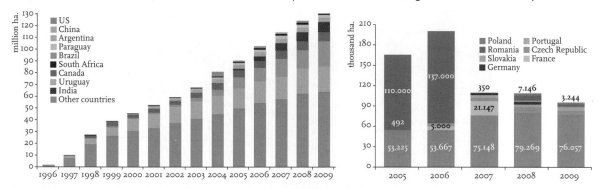

Figure 2.1. Area occupied by transgenic crop, globally (a) and in Europe (b)[7].

[5] www.ey.com/Publication/vwLUAssets/Global_Biotechnology_Report_2007/$FILE/BeyondBorders2007.pdf
[6] www.isaaa.org/resources/publications/briefs/41/executivesummary/default.asp
[7] www.lisconsult.nl/images/stories/Downloads/arealen%20transgene%20gewassen%201996%20-%202009.pdf

It is vital that the new advances are put into practice in a sensible and controlled manner. Gaining the trust of the wider public, by providing them with objective information, has to be the first step. Knowledge of history is useful too. In this chapter we will take a bird's eye view of the first 35 years of modern biotechnology. The pertinent legislation is also briefly discussed.

2.1. THE HISTORY IN A NUTSHELL

Biotechnology is older than documented history. To give an example, malting and brewing were already taking place in Mesopotamia (current-day Iraq) in 4000 BC. In China a description of mould growth on grain for the production of alcoholic beverages and vinegar can be found in a book by Confucius (500 BC). However, what we currently refer to as modern biotechnology originated millennia later, i.e. in the second half of the last century. As mentioned in Chapter 1, scientists Paul Berg, Stanley Cohen, Herbert Boyer and their co-workers conducted the first successful recombinant DNA experiments in California in 1973 (Textbox 2.1), thus heralding the advent of modern biotechnology. In so doing, they ever so slightly lifted the lid of a new Pandora's box. In order to ensure that only good things would emerge this time, they continued with the research only after the guidelines for safe experimentation had been established. When this happened, two years later, the research progressed in leaps and bounds on a global scale.

STONE AGE BEER DELIVERY

TEXTBOX 2.1.

The recombinant DNA technology.

The
ACGT
of
life

In recombinant DNA (rDNA) technology, changes are made to the genetic make-up of an organism. Plasmids play an important role here. A plasmid is a stable, lone, usually circular piece of DNA, which can self-replicate in a host cell. It is therefore not part of the genome of a cell. Plasmids are well-known for being able to transfer resistance to antibiotics, because they can easily pass from one cell to another. And that is exactly what Boyer and Cohen did with their first successful rDNA experiments[8]. They started from a plasmid with a gene that caused resistance to the antibiotic tetracycline. Using molecular "cut-and-paste" work they added a new gene encoding resistance to the antibiotic kanamycin to this plasmid. They let this recombinant plasmid be taken up by E. coli. These bacteria later appeared to become resistant to both tetracycline and kanamycin. Both genes were therefore expressed by the "recombinant" cells, proof that their experiment had succeeded. This was done in 1973, the year regarded as the dawn of modern biotechnology. Since then, all kinds of techniques have been discovered, for example, activating and deactivating (silencing) genes. The end is not in sight and the number of areas for application seems inexhaustible.

As described in Chapter 1, the first commercial applications of modern biotechnology followed less than a decade later in 1982. The Dutch company Intervet was first in line with a recombinant vaccine against swine diarrhoea. They were closely followed by the American company Eli Lilly, which manufactured human insulin by using genetically modified bacteria. Since then, the genetic modification of microorganisms (bacteria, yeasts, moulds) has become routine, and has resulted in a whole range of new spin-off products on the market, as we shall see later. A further ten years later, at the end of the '80s and beginning of the '90s, genetically modified, or transgenic, animals and plants came on the scene. The Dutch bull "Herman" and the American "Flavr Savr" tomato were the trendsetters in

THE SIXTIES AND SEVENTIES WERE PRETTY TOUGH FOR PIG BREEDERS
CRAP, MAN!!!

[8] web.mit.edu/invent/iow/boyercohen.html

TEXTBOX 2.2.

Herman, the transgenic bull.

The Dutch firm Pharming is the 'creator' of the late bull Herman, a 'blaarkop' (breed of dairy cattle) to whose genetic material a human gene was added. When Herman was still an embryo, scientists at Pharming added the human gene that encoded lactoferrin to the DNA in his cells. Human lactoferrin is an infection-inhibiting protein that occurs in substantial quantities in mother's milk. The immune system of babies is not yet sufficiently developed and lactoferrin helps protect them against infection. The human lactoferrin gene was synthesised by researchers at Pharming and inserted into the cells of the Herman embryo. This was done in the hope that milk from Herman's female offspring would contain substantial quantities of this substance. The intention was to extract lactoferrin from the milk and market it as a drug for people with a compromised immune system, e.g. AIDS patients. The announcement about it being added also to food for premature babies generated a lot of fuss (Figure 2.3). Initially, the reasoning was that Herman's female offspring would be better protected against mastitis (udder infection), and so less antibiotic would be needed to maintain the health of these transgenic offspring, and there would be a resulting reduction in antibiotics in the food chain. After many years in the headlines, Herman was put to sleep on Friday 2 April 2004 by doctors from the Faculty of Animal Health in Utrecht. The board of the Herman Bull Foundation decided on euthanasia because the animal was suffering too much from old-age arthritis. The bull, the first genetically modified cattle in the world, was 13 years old. Herman's skin is now on display in the Leiden museum, Naturalis. On the Pharming website[9] you can read that the company is close to marketing human lactoferrin as an advanced ingredient in nutritional products.

---

this area (Textbox 2.2 and 2.3).

Meanwhile, the number of modern biotechnology applications has multiplied, with a great many still in the pipeline. Biotechnology companies and institutes have introduced new drugs, vaccines, diagnostic tests, medical treatments, environmentally friendly products and foodstuffs. One of the most spectacular developments has been the cloning of adult mammals; "Dolly the sheep" being the now legendary example of this technology (Textbox 2.4).

In addition, genetically modified variants of a great many crops have been created. These modifications are intended to make the varieties resistant to disease or certain pesticides, or sometimes to allow extra food to be produced, or to enable them to grow in poor conditions. Golden Rice is an oft-discussed example. It is a recombinant variant of rice that contains beta carotene, a substance converted to vitamin A in the body. Vitamin A is essential for a healthy immune system and good eyesight, amongst other things. This crop is intended to compensate for a shortage of vitamin A in the diet of those living in developing countries. This shortage has until now

[9] www.pharming.com

Figure 2.2. Mother's milk from cows. Controversial poster issued by the Dutch League for Animal Protection in 1994. We thank this league for providing archived material and the approval to use it for publication.

made many people needlessly blind. Meanwhile a new recombinant variant has been developed that contains up to 23 times more beta carotene than the original transgenic variant. More about Golden Rice in Chapters 3 & 8.

Plant breeding stations have since grown an entire range of genetically modified plant varieties and brought them onto the market. In this way corn varieties (the so-called Bt crops) have been made resistant to the European corn borer (*Ostrinia nubilalis*), with the introduction of a gene that codes for a protein toxin of the soil-dwelling bacterium *Bacillus thuringiensis*; and varieties of corn, cotton, rapeseed and sugar beet (the H(erbicide) T(olerant) crops) have been made invulnerable to specific herbicides. As a result these crops can now be sprayed with those specific herbicides, for example, Roundup Ready, allowing for the destruction of competing weeds but not the plants that are to be harvested. Research carried out at the University of Illinois shows that growing these herbicide-resistant crops is better for the environment than using other techniques for getting rid of weeds (*AgraFood Biotech*, 23 July 2007). The US is ahead of the game in the introduction of these transgenic crops. As an illustration, more than 90% of the soya bean crop in that country consists of HT crops. In the same issue of *AgraFood Biotech* there is also an article expressing the expectation that, faced with pressure from fast-rising food prices, acceptance in the European Union (EU) is also set to increase rapidly. This expectation was partly based on the increase in the production of transgenic crops in the EU in 2007, mainly in Spain and France. However, since then only a decrease has been witnessed and the areal coverage is extremely marginal in size compared to countries like the United States, Argentina, Brazil, India and China. It remains a very difficult issue in the EU. However, as mentioned above, in mid-2010 it seems that the tide is starting to turn.

TEXTBOX 2.3.
The transgenic tomato "Flavr Savr".

On 18 May 1994 the FDA declared the transgenic tomato Flavr Savr from the firm Calgene (now Monsanto) to be as safe as a traditional tomato. This signalled a breakthrough for the first commercial food product made by using recombinant DNA technology. By inserting an extra piece of DNA, Flavr Savr was genetically modified so that a specific gene is not expressed, i.e. the protein encoded by this gene is not made by the cell. This process is called antisense technology. The protein in this case was polygalacturonase (PG), an enzyme that accelerates the rotting process. PG breaks down the pectin in the cell walls, so that the tomatoes become soft and rot faster. This genetic modification extends the shelf life of the tomato by at least ten days. It also allows the tomatoes to ripen fully and turn red on the plant. This is in contrast to most traditional supermarket tomatoes, which are plucked when green and then treated with ethylene gas (especially in the US) to turn them red, but which have no time to develop taste and aroma.

2.2. SUPPORTERS AND OPPONENTS

At the end of the last century the expectation that modern biotechnology would revolutionise agriculture, health care and environmental protection was so great that, in a cover story in Business Week magazine of 27 July 1998, the 21st century was hailed as the century of biotechnology[10]. The introduction of such

Flavr Savr is therefore, as the name would suggest, a more flavoursome tomato. In 1995 Calgene received authorisation to sell Flavr Savr in Canada and Mexico. The tomato was sold for a number of years in about 3000 shops under the name MacGregor. It has since been taken off the market because the yields were too low. You can read everything there is to know about Flavr Savr in a book by Belinda Martineau, one of the researchers at Calgene who worked on the development of Flavr Savr. The book is called "First Fruit. The creation of the Flavr Savr tomato and the birth of biotech food" (Martineau, 2001).

'Flavr Savr' Tomato

Traditional Tomato

[10] www.businessweek.com/1998/30/b3588002.htm

a spectacular and revolutionary new technology is obviously not without its critics. As already mentioned, the UK's Prince Charles is a very public opponent. In the '90s he set up a website to stimulate debate between supporters and opponents. Prince Charles himself said the following on the subject: "Genetic engineering takes mankind into realms that belong to God and to God alone."

SCIENCE COMPETES WITH RELIGION OVER GENETIC ENGINEERING

There is even opposition in the US, where modern biotechnology has been most widely implemented and most readily accepted by society. As a result, in May 1998 a coalition of scientists, consumer organisations and religious groups filed a lawsuit against the FDA with the intention of banning 36 genetically modified foodstuffs from the market, and of forcing the FDA to test these products more thoroughly and provide them with a label before re-releasing them onto the market. In October 2000 the judge declared in favour of the FDA, which was a victory for the American biotechnology camp and evidence of the stark contrast in the way we deal with this issue in Europe. The standpoint of the Bush government was, however, far from unambiguous. On the one hand Bush wanted to incorporate the development of biotechnology in a Bio-Shield project[11] to combat potential bioterror, while on the other hand he was attempting to introduce a total ban on stem cell experiments; his successor Barack Obama has a completely different view on the latter (see Chapter 14). The policy of the Bush government was intended, out of economic necessity, to vigorously stimulate the export of genetically modified agricultural products to Europe. All this makes it clear that, for the time being, a continuing social debate on where we want to draw the line in the matter of modern biotechnology is a must. It is unrealistic to suppose that progress in this field can be halted, given the speed with which advances and developments in biotechnology have occurred in countries like the US, Argentina, Mexico and China. It's vital to put this novel technology into practice in a sensible and controlled manner to prove safety and usefulness beyond any doubt. To be more specific, we should now put our knowledge and experience for instance into the area of high-tech agriculture for the production of health-promoting foodstuffs and

[11] www.hhs.gov/aspr/barda/bioshield/index.html

high-grade products such as medicines by using genetically modified (transgenic) plants and animals, initially under well-controlled conditions in high-tech glasshouses or stalls. An interesting example is the cultivation of transgenic tobacco plants in greenhouses for the production of anti-bodies to treat infections caused by the West Nile virus (Lai *et al.*, 2010). When sufficient trust in safety and usefulness is gained, a wealth of opportunities grows on every bush ready to be explored.

We, the authors of this book, are afraid that the wrong issues are still being debated. Take the speed with which whole genomes are mapped today. As a result of that the individual human DNA passport is now a little too close for comfort. Is that what we want? And if it happens, who gets to approve it, who is allowed to use it? Will it have consequences for our ability to get insurance cover? Will it turn our health-care system upside down? Will it have far-reaching consequences for our criminal code? These are examples of questions that we should now be debating publicly so as to lay down the limits of what we want in advance. Not later, when it becomes nigh on impossible to reverse developments in progress.

One admirable first move in the Netherlands to raise an emotionally charged topic at an early stage was made in *Trendanalysis Biotechnology 2007 Chances and Choices* (a pdf of the full report in Dutch can be downloaded[12]). The memorandum looked at ethnicity as a possible factor in scientific research, genetic diagnostics and genetic population studies. The conclusion stated: "The genotype for certain disorders differs from one ethnic group to another. The efficacy of the drug treatment for diseases is also affected by the patient's genetic background. However, ethnicity is an emotionally charged issue in Europe, especially where genetics is concerned. At present the ethnic background of patients and clinical trial subjects is not automatically registered in the Netherlands. The absence of this information hinders genetic diagnosis, population studies and scientific research. The government must decide whether or not to allow the registration of ethnicity, and if so, for what purpose and under what conditions." Well-distributed information and well-prepared public debates are, we believe, indispensable in this context.

It is not just the Netherlands that is struggling with these revolutionary developments. Almost ten years after the controversial poster issued by the

[12] www.cogem.net/ContentFiles/Trendanalyse%202007.pdf

MAdGE milking the subject

MAdGE (*Mothers Against Genetic Engineering in Food and the Environment*) have launched a highly controversial billboard campaign in Auckland and Wellington which is, they say, designed to provoke public debate about the social and cultural ethics of genetic modification in New Zealand.

The billboards depict a naked, genetically engineered woman with four breasts being milked by a milking machine. The woman has GE branded on her rump.

Source: Getty Images

Dutch League for Animal Protection (Figure 2.2), we found the extract and poster below in the October 2003 issue of *AgraFood Biotech*. It seems that such provocative campaigns are needed to wake up the general public and get them involved in such far-reaching developments. The discussions about Herman have, in any case, helped result in extensive legislation in the Netherlands. We will return to this later in the chapter. A scientific reconstruction of the public debate on Herman can be found in the PhD thesis, *A calf is born*, by Elmar Theune (2001).

2.3. WHY TRANSGENIC PLANTS?

Plants are genetically modified for four application-oriented reasons:

1. To improve taste, shelf life and/or dietary value, as in the case of Flavr Savr, and to obtain products that prevent disease and disorders and have healing properties (novel and functional foods; nutraceuticals). The science in this area is still in its infancy, but there are nevertheless huge expectations for the long term. Transgenic rice and

sorghum are examples of existing products, which will be discussed in Chapter 8.

2. With the aim of making a plant easier and less expensive to cultivate. Disease and pest resistance, protection against cold, drought and/or frost, immunity to herbicides, these are the properties that scientists want to impart to plants. Transgenic soya, corn, cotton and canola (rapeseed) are examples of transgenic crops that are already being produced on a large scale. The primary aims in this first generation of transgenic plants are disease and pest resistance and immunity to herbicides. A newer example is the transgenic tomato which grows very well in salty soil. The eminent journal *Nature Biotechnology* published an article by Zhang and Blumwald (2001) in which they describe how they had created genetically modified tomato plants that not only flourish but produce edible fruits on extremely salty soils, because the tomato fruit doesn't take up salt from the soil. According to the researchers their discovery means that 60 million hectares of land around the world will immediately become available for the cultivation of such transgenic, salt-tolerant crops; land where previously not much would grow. A promising approach to combating the notorious potato late blight *Phytophthora infestans* comes from Wageningen University[13]. By incorporating combinations of resistance genes (R-genes) from resistant wild potatoes into existing commercial potato varieties (cisgenesis), researchers at Wageningen believe that they can protect these potato varieties sustainably against this disease. The

idea is to insert an optimum combination of three or four resistance genes into a so-called 'gene cassette' and to incorporate these into a potato variety. Figure 2.3 is a photo of one of their field trials. By producing more of these cassettes with varying combinations of R-genes, it is possible to make a variety flexible and therefore resistant in the long term. It is possible to make variations in space or time, as happens naturally in the wild family strains, by initially working for a few years with one gene cassette while keeping the other in reserve, or by using them in combination.

Figure 2.3. Original potato variety Désirée is diseased and surrounded by healthy Désirée plants supplied with a Phytophthora resistance gene (thanks to the DuRPh researchers[14] for the photo).

According to the researchers this method of genetic modification makes it possible to create sustainably resistant varieties of much loved classics, such as the more than one hundred year old Bintje or Russet Burbank. It is precisely this that makes the approach so elegant. The researchers are also devising

[13] www.cisgenesis.com

[14] www.durph.wur.nl/UK

strategies for this resistance management. The question is whether it can also be used in the short term. Genetically modified potatoes for consumption are still taboo within Europe and so for the moment do not get further than the glasshouses and trial fields, thus still far from our tables. The researchers argue, however, that the cisgenic crops only contain genes that can be used in classical breeding too. They argue that cisgenesis is as safe as or even safer than classical breeding and propose a faster and cheaper approval of cisgenic crops.

3. For obtaining plants that produce pharmaceutical and other high-value substances. Although plants have long been the source of many raw materials for the pharmaceutical industry (a quarter of medicines contain substances of plant origin), this terrain is still virtually barren as regards transgenic plants, especially compared to transgenic animals. There is, however, enormous potential, partly because diseases (such as BSE) that can be transmitted to humans are much less of a problem in the production of plants. Since the mid '90s an increasing number of articles have outlined the possibilities of this so-called molecular pharming with transgenic plants and the market is fast approaching. Following hundreds of field trials in the United States and Canada dozens of medicines manufactured by using transgenic plants have already entered various phases of clinical trials, including several in a very advanced phase. Five of them, including human lactoferrin and lysozyme from rice, are being produced commercially on a small scale for use as chemicals. In recent years there has also been more focus on the legislation, particularly concerning the manufacture of pharmaceuticals from transgenic plants (Spök *et al.*, 2008). Ramessar *et al.* (2008) are looking at these things in detail for transgenic corn as a production platform.

4. For obtaining plants as alternative feedstocks for the production of biofuels. At CHI's 2009 Advanced Biofuels Development Summit there was a consensus that members of the biofuel industry are ready to meet the challenge of producing replacements for petrochemical fuels that will be cost-competitive and renewable, and that will meet the increasingly stringent demands of the Green Revolution[15]. Nowadays, we indeed see that there are huge investments in the shift from an oil-based to a biomass-based society. Today we are also experiencing the debate on the question *Food or Fuel?*, which we endorse. However, we leave it with this statement and focus solely on food and health in this book. Biofuels are books in themselves (e.g. Vertes *et al.*, 2009).

The long-term opportunities and possibilities for small, high-tech countries like the Netherlands are the development and testing of new technologies, especially for items 1, 2 and 4, because in this respect the current 'farming' methods will not differ greatly from those required for transgenic crops. As regards item 3, some things are a little different. The agricultural acreage required for this is small, but the technology

---

[15] www.genengnews.com/gen-articles/alternative-feedstocks-boost-ethanol-production/2942/

is advanced. That is where the principle development opportunities and production possibilities lie for countries like the Netherlands.

PLANTS ARE A GREAT RESOURCE FOR THE PHARMACEUTICAL INDUSTRY

SI SENHOR, OF COURSE... ...WE'RE PHARMACISTOS!

2.4. WHY TRANSGENIC ANIMALS?

Similar in some aspects to plants, animals are genetically modified for the three following application-oriented reasons:

1. To render them suitable for the production of high-purity (human) proteins, in particular medicines. Dozens of medicines made from transgenic animals are currently in various stages of clinical trials. For commercial applications and depending on the product, herd sizes must be several dozen (Factor IX - a drug for haemophilia B - made from transgenic sheep or pig's milk) up to several thousand (α-anti-trypsin - a drug for cystic fibrosis - made from transgenic sheep or goat's milk). The conventional reproduction methods require a lot of time and money and there is also a good chance that the foreign gene gets lost in the process. The requirement for efficient reproductive cloning techniques with genetically identical offspring is particularly high in relation to larger herds. The quest for this has already resulted in Dolly the sheep (Textbox 2.4), the first cloned mammal. She was made from an adult udder cell and died on 13 February 2003 from ailments resulting from premature ageing.

2. To make them more suitable as organ donors for xenotransplantation (see also Chapter 12). Xenotransplantation is the transplantation of animal organs to humans. American scientists have transplanted insulin-producing cells from pigs into monkeys with type 1 diabetes. The subsequent average survival of these monkeys was six months, without insulin injections and with normal blood sugar levels. In the March 2006 edition of *Nature Medicine* plans were announced to conduct this research on humans before 2010 (Hering *et al.*, 2006). However, a neck-and-neck race between xenotransplantation and stem cell therapy threatens to emerge (see also following paragraph and Chapters 12 and 14). Political policies and social acceptance, as well as

technological advances, will determine who wins the race in the end, or whether both will come in equal first.

3. To obtain "better" farm animals and fish. As an illustration of this, work is being done on improving disease and plague resistance, on pigs that can digest cellulose, pigs with leaner meat, sheep that can use low-cysteine feed, frost-resistant salmon (with anti-freeze genes from cold-resistant fish from the Arctic and Antarctic oceans) and giant salmon. An example of the latter is the AquAdvantage salmon, made by inserting a gene from an eel and the growth hormone gene of a specific salmon into the Atlantic salmon, causing this genetically modified salmon to grow twice as fast.

SALMON INDUSTRY THRIVES THANKS TO AQUADVANTAGE SALMON

Suggestions that it will grow six times the size of a normal salmon are pure fantasy: the final size is the same. However, possibly the most significant advantage for the industry is that the AquAdvantage salmon will also continue to grow in cold conditions. This is in contrast to the non-modified variety and means that modified salmon can be harvested twice a year instead of the usual once. They also need 30% less food to reach their harvesting weight. AquaBounty Technologies, the producer, expects this salmon to be on the market by 2010[16].

TEXTBOX 2.4.
Dolly the clone.

"On 5 July 1996 the most famous lamb in history entered the world, head and forelegs first. No one broke open champagne. No one took pictures. Yet the birth of this remarkable creature was soon to provoke amazement and wonder around the world. Created in Edinburgh's Roslin Institute by embryologist Ian Wilmut, Dolly was born not from the union of a sperm and an egg, but from the genetic material of a single cell taken from a six-year-old sheep. Dolly is the first creature to be cloned from an adult cell in this way – an achievement which for years leading scientists deemed biologically impossible." This is a literal quote from the inside of the jacket of Gina Kolata's book "Clone: the road to Dolly and the path ahead", a very worthwhile read, published in 1998, about the history of Dolly and all the events surrounding it.

Dolly was created from a so-called specialised cell, in this instance an udder cell, from a six-year-old sheep. Because DNA undergoes "wear" with every cell division, Dolly emerged from relatively old genetic

[16] www.aquabounty.com

material, immediately giving rise to the question of how old Dolly actually was at birth. Three years later it was revealed that Dolly had aged relatively quickly and in 2003 she died of early ailments of old age. In theory this problem can be prevented by cloning with

unspecialised or stem cells (diagram below), which can for example be isolated from embryos, umbilical cord or bone marrow. This is one of the reasons why stem cell research is the focus of so much interest (Section 2.5 and Chapter 13).

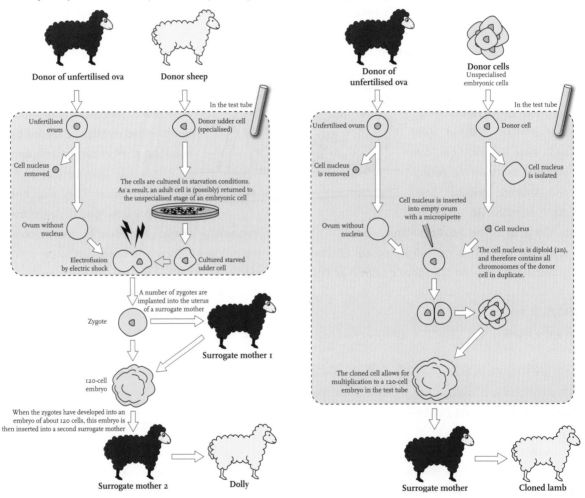

The diagrams are adapted versions of the ones originally drawn up by Jos van den Broek, currently Extraordinary Professor in Biomedical Scientific Communication (Leiden University, the Netherlands).

There are a multitude of long-term opportunities and possibilities in all these areas. The applications in 1 and 2 require highly specialised knowledge and experience, not only of the molecular biology of animals, but also of industrial high-tech farms, while only a small amount of land is required. In short, an ideal starting point for countries like the Netherlands to develop and test new technologies, and to set up new production projects with transgenic (cloned) animals on high-tech farms. The Dutch company Pharming, the "creator" of Herman and a global leader in this area, is a good example of this. However, the introduction of an animal welfare act, which bans nuclear transplantation (see also Chapter 14) - the cloning technique used by Pharming - together with economic and epidemiological reasons, has forced this company to look abroad, at least as far as production is concerned. Thus, the Dutch legislation is now restricting opportunities. This is also a conclusion in the latest *Trendanalysis Biotechnology 2009*. In this Dutch memorandum commissioned by the Dutch government, one of the biotechnology dilemmas for policy and politics reads: "Cloning of animals is not allowed in the Netherlands, but an import ban on cloned animals, descendants or products of cloned animals is not sustainable; the government faces the choice to accept import or to adapt legislation."

2.5. GENE AND STEM CELL THERAPY

Gene and stem cell therapy are biomedical developments which have already caused a lot of controversy and about which opinions differ widely. On paper they look like a panacea for almost every disease, but the practical obstacles that still need to be negotiated are many and complex and constitute a major challenge for biotechnological researchers. We will briefly look at them here and come back to them in more detail in later chapters.

BIOMEDICAL THERAPY HAS LONG SINCE MOVED ON FROM THE EXPERIMENTAL STAGE

WHOAH! THIS SHOULDN'T HAPPEN ANYMORE!

GRRR

Gene therapy is a treatment for curing congenital disorders. Congenital diseases arise because of an abnormal or missing gene that causes disease symptoms. Gene therapy involves inserting a healthy gene into the cells of the patient. It isn't always necessary to do this in all cells. That would effectively make gene therapy unviable. Around the world there are only a few very specific and rare cases in which gene therapy has been at all successful in humans. Unfortunately there are also cases in which the patient has died as a result of the treatment. Chapter 11 contains a more extensive discussion of the positive and negative aspects of gene therapy.

In contrast to gene therapy, in at least one application, stem cell therapy has long since moved on from the experimental stage. Stem cells from bone marrow have been used for more than 30 years in the treatment of some forms of cancer, for example leukaemia. Chemotherapy kills not only cancer cells but also most other growing cells, including the stem cells in bone marrow. Which is why bone marrow is taken from the patient before chemotherapy is started. Where possible, healthy stem cells are isolated, then stored and injected into the patient after the therapy, to provide more healthy growing (blood) cells. If this doesn't work, bone marrow from a donor is used.

Roughly speaking there are two sorts of stem cells: embryonic stem cells and adult stem cells. As the name suggests, the former are taken from embryos. When they are isolated from embryos that have only undergone a limited number of cell divisions, they can grow into any kind of body cell. Adult stem cells are isolated from bone marrow, umbilical cords, etc., and are already slightly differentiated. They can only develop into a limited number of related cell types, for example, blood stem cells into blood cells such as red and white blood cells.

In 1998 a method was published for the first time to cultivate human embryonic stem cells *in vitro*, i.e. outside of the body in test tubes or Petri dishes. This caused a great deal of excitement in the scientific world, with visions of a revolution in health care, but considerable unrest among religious and bio-ethic groups and public policy-makers. The latter protested primarily against the 'misuse' of human embryos. However, the real stem cell hype occurred in April 2001, when the American biologist Daniel Orlic (Orlic *et al.*, 2001) wrote an article in the leading scientific journal *Nature* describing how he had largely repaired a mouse heart, damaged following a heart attack, by injecting the animal with bone marrow cells. The cells appeared to develop into heart cells, dramatically improving the functioning of this organ, but the exact role of the cells in the repair remained unclear.

"The principle of stem cell therapy is as simple as it is fantastical", said stem cell expert Christine Mummery in the January 2007 issue of a Dutch newspaper (*NRC*). "Stem cells can develop into specialised tissue and thus replace diseased tissue. So, new heart muscle tissue could be created after a heart attack, or Parkinson's disease could be cured by creating new brain cells. But after years of research we still have no idea what really happens when stem cells are introduced." In the same year she also stated that stem cell research was advancing slowly and with varying success mainly because President George Bush had banned the use of government money for embryonic stem cell research. She believed that the EU was also holding back subsidies for stem cell research. All this basically means that stem cells have far from fulfilled their potential. "Stem cell therapy in this form is still in a research phase, so treatments outside the application of scientific research are banned", said the Dutch Ministry for Health in early 2007. In so saying, the minister answered the call to stop stem cell pirates made a year earlier by various medical specialists and stem cell experts. Unlike in Spain, Belgium and the United Kingdom, commercial stem cell therapy had been permitted until then in the Netherlands, and for several years already two Dutch companies had been treating hundreds of incurably sick patients suffering from, for example, a spinal cord lesion or multiple sclerosis, with expensive, unproven and unauthorised stem cell therapies.

STEM CELL PIRACY

This ministerial decision ended that treatment. Only university hospitals and the Dutch Cancer Institute were granted a permit to conduct stem cell research. Furthermore, permits are only issued if the Central Committee on Research involving Human Subjects gives approval. This judgement puts the focus firmly on safety, detailed background studies, accurate research design and comprehensive information to patients - and rightly so. At least in the US, however, a fair wind has begun to blow for stem cell research since the arrival of Bush's successor, Barack Obama, in January 2009; more about this in Chapter 14.

2.6. EU LEGISLATION

In the EU two provisions concerning legislation on products involving modern biotechnology came into force in 2003 and are now being implemented: one concerning genetically modified foodstuffs and animal feed, and one concerning the traceability and labelling of genetically modified organisms and the traceability of foodstuffs and animal feed produced with genetically modified organisms. The cornerstone of the EU risk analysis regarding safety in food and animal feed is the European Food Safety Authority (EFSA). In close collaboration with national authorities and in open consultation with all stakeholders, the EFSA gives independent advice and makes sure there is clear communication on existing and expected risks. Figure 2.3 gives an overview of the EU approval procedure for food and feed originating from genetically modified plants. The guidelines followed in this process are open to the public and can be downloaded from the EFSA website[17].

The legislation does, however, differ between EU member states and non-EU countries, but the safety assessments in the various national procedures are nevertheless based on the internationally recognised consensus approach of comparative safety analysis (substantial equivalence). This approach involves comparing the transgenic crop in question with a conventional equivalent with a history of safe use. This comparison looks, for example, at the phenotype (the external appearance of an organism, determined by the interaction of the genetic properties – the genotype - and the environment) and chemical composition. Further safety tests are carried out on the basis of perceived differences. The usual items in this risk analysis are the possible presence of toxins and allergens, the risk of transferring the imported genes, and unintentional effects as a result of the genetic modification.

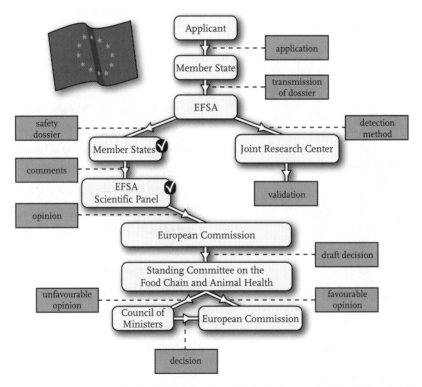

Figure 2.3. Overview of EU approval procedure for GM food and feed (ticks indicate points at which safety is assessed; thanks to Gijs Kleter for the overview, RIKILT Wageningen, the Netherlands).

[17] www.efsa.europa.eu

The second provision concerning traceability and labelling requires all foodstuffs and foodstuff ingredients originating from genetically modified crops, including those without perceivable transgenic material, to be labelled. In addition, there must be a documented system of traceability. Labelling is not required if the products originate from conventional crops, where unavoidable combination with the transgenic equivalent is less than 0.9 per cent. In some member states, for example the Netherlands and Germany, GMO-free labelling (i.e. prepared without gene technology) is permitted. But the criteria here are even stricter (Kleter & Kuiper, 2006).

2.7. CONCLUSION

In 2009 we carried out an analysis of the controversies provoked by modern biotechnology. This analysis is based on a long list of articles that, almost without exception, were published after the turn of the century in renowned scientific journals. Many of them were used in the writing of this chapter. Using this analysis we have come to the conclusion that there are still multiple opportunities and possibilities for the EU to develop into a leading transgenic production area. There is also a firm basis for researching standards and norms for the safe implementation of these new technologies, without any unacceptable risk to health and the environment. A lot depends on future EU-governments policies in this area as well as social acceptance. Science and industry are ready. The farmers are generally well-trained, have the necessary experience and are looking for alternatives. In short: it's time for the EU to seize the day and become a leading transgenic production continent. In the next chapter we propose seven points that need attention to remove the EU hesitation toward gene technology in agriculture. The website GMO Compass[18] has been created using EU money and is a very good source of information on new developments in modern biotechnology.

[18] www.gmo-compass.org/eng/home

2.8. SOURCES

Hering, B. J., Wijkstrom, M., Graham, M. L., Harstedt, M., Aasheim, T. C., Jie, T., *et al.* (2006). Prolonged diabetes reversal after intraportal xenotransplantation of wild-type porcine islets in immunosuppressed nonhuman primates. *Nature Medicine, 12*(3), 301-303.

Kleter, G. A., & Kuiper, H. A. (2006). Regulation and risk assessment of biotech food crops. In P. K. Jaiwal & R. P. Singh (Eds.), Metabolic engineering and molecular farming II, Plant genetic engineering (Vol. 8, pp. 311-337). Houston TX, Studium Press.

Lai, H. F., Engle, M., Fuchs, A., Keller, T., Johnson, S., Gorlatov, S., *et al.* (2010). Monoclonal antibody produced in plants efficiently treats West Nile virus infection in mice. *Proceedings of the National Academy of Sciences of the United States of America*, 107(6), 2419-2424.

Martineau, B. (2001). First fruit: The creation of the Flavr Savr Tomato and the birth of biotech foods. McGraw-Hill Companies.

Orlic, D., Kajstura, J., Chimenti, S., Jakoniuk, I., Anderson, S., Li, B., *et al.* (2001). Bone marrow cells regenerate infarcted myocardium. *Nature, 410*(6829), 701-705.

Ramessar, K., Sabalza, M., Capell, T., & Christou, P. (2008). Maize plants: an ideal production platform for effective and safe molecular pharming. *Plant Science, 174*(4), 409-419.

Spök, A., Twyman, R., Fischer, R., Ma, J., & Sparrow, P. (2008). Evolution of a regulatory framework for pharmaceuticals derived from genetically modified plants. *Trends in Biotechnology, 26*(9), 506-517.

Theune, E. (2001). A calf is born: a reconstruction of the public debate on animal biotechnology. PhD Thesis, Wageningen.

Vertes, A., Qureshi, N., & Yukawa, H. (2009). Biomass to biofuels: strategies for global industries, Wiley.

Zhang, H., & Blumwald, E. (2001). Transgenic salt-tolerant tomato plants accumulate salt in foliage but not in fruit. *Nature Biotechnology, 19*(8), 765-768.

GENETICALLY MODIFIED CROPS AND THE EUROPEAN UNION

Outside the European Union (EU), the area planted with genetically modified crops (GM crops) increases about 10% annually (see Figure 2.1 in preceding chapter). Within the EU there are still seemingly unbridgeable differences in opinion and acceptance among the Member States. The EU regulation to approve GM crops is very restrictive. By analysing the controversial issues, especially concerning food and environmental safety, we arrive at seven points that need attention to remove EU hesitation towards gene technology in agriculture (Table 3.1). We have included this chapter especially for the policymakers, but we hope that it will interest the layman as well. This chapter is complementary to Chapter 8.

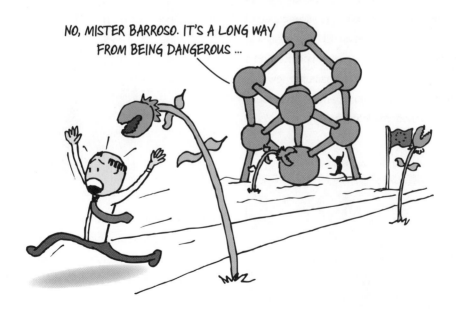

Global food supply is increasingly under pressure as a result of the ever-expanding world population and the growing welfare in developing countries. The transition from fossil-based processes to biomass-based production, in particular fuels, further aggravates this (Boddiger, 2007; Gressel, 2009). In the future, agriculture will no longer focus solely on the production of food and feed, but also on fuels, chemicals and pharmaceuticals. This will further increase the pressure on the food supply. Drought and flood also threaten a sustainable food supply. To deal with all these challenges, full input from modern biotechnology, in particular gene technology is in our opinion an inevitable and essential measure. However, the issue of whether or not to adopt genetically modified crops (GM crops) is still a cause of heated debate in the European Union (EU). By analysing the controversial issues we come to seven points (Table 3.1) that need attention to remove hesitation towards gene technology in EU agriculture, and to eventually obtain a firm 'yes' to modern biotechnology.

Table 3.1.
Seven focal points to accept GM crops within the EU.

1. Long-term studies to monitor the safety of GM food/crops.
2. Do not polarize!
3. Global uniformity.
4. Risk assessment.
5. Development of a SMART legislative framework.
6. Responsible progress hand in hand with ongoing public debate.
7. Integrated approaches for Third World countries.

To convince all EU members to say yes, many EU consumers and in particular politicians, must shift their attitudes toward GM crops. The first step is to recognise the advantages; a step taken for many years already by the Vatican (Textbox 3.1). Convincing examples of GM crops that result in environmental protection, better land use, phytoremediation (the use of (GM)plants to decontaminate polluted water and soils), and especially health benefits for individuals, will stimulate this shift and step.

THE VATICAN APPROVES GENETIC MODIFICATION

For the main GM crops, i.e. herbicide-resistant/tolerant and insect-resistant crops, positive farm and environmental impacts have been proven. Strategies for better land-use are now developed by breeding new crop varieties resistant to drought, salinity and other extreme environmental stress. Removal of cadmium from the soil by GM tobacco is a promising example of phytoremediation (Abhilash, Jamil, & Singh, 2009; James & Strand, 2009; Macek *et al.*, 2008). The publication on anthocyanin-enriched GM tomatoes is an example of a health benefit (Butelli *et al.*, 2008). However, an equally momentous achievement of genetic modification of plants, the Golden Rice developed a decade ago, has been largely ignored, because intransigent opposition by anti-biotechnology activists makes risk-averse regulators adopt an over-precautionary approach that stalls approvals (Miller, 2009). Fedoroff (2008) argues that a new Green Revolution demands a global commitment to creating a modern agricultural infrastructure everywhere and poses the question whether we have the will and the wisdom to make it happen. Miller (2009): "The Golden-Rice story makes it clear that the answer is, not yet." In the following sections we discuss the two main concerns raised by the opponents of GM crops, i.e. food and environmental safety. Our suggested seven attention points (Table 3.1) focus on these two safety issues.

TEXTBOX 3.1.
Vatican: GM not against God's will.

This is the title of an announcement in AgraFood Biotech No 19, December 8, 1999. In October 1999 members of the Pontifical Academy for Life presented two volumes of documents on ethics and genetic technology, the result of more than two years of discussion and study. The documents voiced a prudent yes to GM plants, as the academy was increasingly encouraged that the advantages outweigh the risks. Risks should be examined thoroughly, but 'without a sense of alarm'. A spokesman of the academy said they do not agree with groups or persons (e.g. Prince Charles), who say GM is against the will of God. Since then the Vatican has had an ongoing debate on this topic. For instance, by the end of the seminar "GMO: threat or hope", convened by the Pontifical Council of Justice and Peace in Rome in November 2004, fellow officials sidestepped taking a definitive stance, insisting that they were still in the preliminary stage of gathering information on this difficult issue, while also maintaining that this technology "should not be abandoned" (Fox, 2004). Five years later Nature Biotechnology headed a brief announcement with: Vatican cheers GM (Meldolesi, 2009). It announced a closed door meeting at the Vatican in May 2009 with leading scientists gathering to discuss a campaign backing agricultural biotechnology. The study week was organised by Potrykus, co-inventor of Golden Rice, on behalf of the Pontifical Academy of Sciences. The Vatican has long been concerned about food security, and advisors from the academy, which has a membership list of the most respected names in 20th-century science, have recognised that plant biotechnology has the potential to benefit the poor.

3.2. SEVEN FOCAL POINTS TO ACCEPT GM CROPS WITHIN THE EU

1. Long-term studies to monitor the safety of GM food/crops

More than a decade of consuming GM foods has demonstrated that they have no direct health risks and that the required safety assessments have worked well. Domingo (2007) concludes though, that relatively little scientific toxicological research has been done on GM crops/foods and that long-term studies are necessary. Dona and Arvanitoyannis (2009) have analysed possible health risks of GM crops. They write that intensive work is currently in progress to thoroughly understand and forecast possible consequences on human, animals, and the environment, but anticipate that many years of careful, independent research are needed to accomplish this. They believe that these long-term studies are especially needed to take away any doubt that may exist in society. Although we do not support the whole paper, we endorse this last view. The first item requiring attention is thus, in our

---

TEXTBOX 3.2.

The 20-year environmental safety record of GM trees.

In a commentary of May 2009 in Nature Biotechnology Strauss et al. (2009) call for more science-based (case-by-case) evaluation of the value and environmental safety of GM trees, which requires field trials. However, the regulatory impediments being erected by governments around the world are making such testing so costly and Byzantine, it is now almost impossible to undertake field trials on GM trees in most countries. One year later, in a letter to the editor of the same journal, Walter et al. (2010) summarise the key published evidence relating to the main environmental concerns surrounding the release of GM trees. On the basis of their analysis of a very large amount of performance and safety data related to GM crops and trees gathered since field trials were first initiated in 1988, they pled for a consideration of the opportunity costs for environmental and social benefits, and not just the risks, in its deliberations of field trials and releases. Their search in publicly accessible databases worldwide revealed >700 field trials with GM trees (including forest trees, fruit trees and woody perennials). None of them has reported any substantive harm to biodiversity, human health or the environment. Few GM tree species have as yet been deployed commercially. A notable exception is Bacillus thuringiensis toxin(Bt)-expressing poplar trees in China. Approximately 1.4 million Bt poplars have been planted in China on an area of ~ 300 – 500 hectares along with conventionally bred varieties to provide refugia to avoid the development of Bt resistance in insects. The trees are grown in an area where economic deployment of poplar was previously impossible due to high insect pressure. GM trees have been successfully established and have successfully resisted insect attack. The oldest trees in the field are now 15 years old. No harm to the environment has been reported.

opinion, a clear definition of how long-term studies should be conducted, and a proposal that these should apply to any novel crop/food product. The latter is in line with what Kok *et al.* (Kok, Keijer, Kleter, & Kuiper, 2008) propose, i.e. to develop a general screening frame for all newly developed plant varieties to select varieties that cannot, on the basis of scientific criteria, be considered as safe as plant varieties that are already on the market. They conclude that the current process of the safety evaluation of GM crops versus conventionally bred plants is not well balanced. And we fully support this view. An interesting case is the safety record of GM trees (Textbox 3.2).

2. Do not polarize!

Scientifically there is indeed no reason to test GM foods more thoroughly than other new food products. The fact that it happens, is not only the result of campaigns by anti-biotechnology organisations. In the early 1980s, under the pretext of environmental protection, some of the largest agricultural chemical companies approached the chief policymakers in President Reagan's administration (1981-1988) with a request to set up more restrictive regulations concerning the acceptance of GM crops (Miller *et al.*, 1997). Their suggestions went considerably further than could be justified by scientific reasoning, but their motives were clear: regulations as market barriers for less powerful competitors such as seed companies and biotech starters. Their success has led to the present overregulation, strongly limiting the introduction of new GM crops, but ironically hitting the multinationals themselves most. Ten years later Miller (2008) fulminated again and condemned the decision taken by two university rectors in Germany to forbid scientists to continue their field trials with GM crops. He accused German universities of protecting their reputations by curtailing the academic freedom of faculty and students in the face of demands and threats from ideological bigots. In another paper Miller and co-authors (Miller, Morandini, & Ammann, 2008) take a sharp stance against the publication policy

---

TEXTBOX 3.3.

Modern biotechnology: scientific victim?

"Primarily outside the scientific community, misapprehensions and misinformation about recombinant DNA-modified (also known as 'genetically modified', or 'GM') plants have generated significant 'pseudo-controversy' over their safety that has resulted in unscientific and excessive regulation (with attendant inflated development costs) and disappointing progress.

But pseudo controversy and sensational claims have originated within the scientific community as well, and even scholarly journals' treatment of the subject has been at times unscientific, one-sided and irresponsible. These shortcomings have helped to perpetuate 'The Big Lie' that recombinant DNA technology applied to agriculture and food production is unproven, unsafe, untested, unregulated and unwanted. Those misconceptions, in turn, have given rise to unwarranted opposition and tortuous, distorted public policy."

of some leading scientific journals. By publishing activists' papers with sensational, inaccurate claims they provoke, according to Miller *et al.* (2008), pseudo-controversy, misapprehensions and misinformation about GM crops, especially concerning environmental or health risks (Textbox 3.3).

We think that Miller and co-authors are right in principle, but the way they express it will not unite the warring parties, and they are not the only ones using forceful language. We believe that, for the time being, too many tests may be preferable to too few, given the ever-present hypersensitivity to GM food and crops among the opposition groups, but also among many policymakers and consumers.

3. Global uniformity

Some opposition groups request that lengthy toxicological tests lasting at least two years should be conducted with GM foods. The European Food Safety Authority (EFSA) reports that 90-day food tests on animals, mainly rats, are usually sufficient to demonstrate the safety of GM foods, provided that these tests are performed according to international guidelines (Konig *et al.*, 2004). The report calls for a more uniform approach to food testing and the use of new (profiling) technologies. It suggests a solid pre-market risk analysis rather than monitoring after a product has been marketed. Risk assessment of predictable effects is easily attained through specific *in vitro* and clinical tests. Some institutions, such as the Food and Agriculture Organization and the World

Health Organization, require an estimate of any unpredictable and unintended effects, even if there is no indication that such effects are more likely to occur in GM crops than in conventional ones (Batista & Oliveira, 2009). There are two different approaches for this purpose (Kuiper, Kok, & Engel, 2003). One is a targeted approach that is regularly used to evaluate new GM foods. Here, several key nutrients are analysed that, if inadvertently altered, could influence the nutritional value and eventually the safety of the modified product. This approach does not consider any unknown anti-nutrients and natural toxins. The second approach is non-targeted and based on profiling methods, in which potential alterations in GM food that occur at the genomic level, as well as at the levels of gene expression, translation and metabolic pathways, are evaluated. Several recent studies have begun to explore profiling methods that aim to increase the probability of detecting any unpredictable, unintended effects and, consequently, improve the efficiency of GM food safety assessment (Batista & Oliveira, 2009). Profiling techniques are a potentially powerful, complementary tool, offering the capacity to broadly screen for possible changes at different integration levels of cells or tissues in a non-selective, impartial manner. For these reasons we have formulated this third attention point.

4. Risk assessment

Environmental safety policy is generally built on the precautionary principle (Textbox 3.4). The precautionary principle, for example, is the basis of EU Directive

2001/18, which states: "In accordance with the precautionary principle, the objective of this Directive is to approximate the laws, regulations and administrative provisions of the Member States and to protect human health and the environment…" For GM crops it can be assumed that pollen from a GM crop will pollinate a wild variety, if they grow in each other's neighbourhood. GM seeds spread with wind or birds, and end up in the wild. It is thus conceivable that the (pollinated wild variety of) GM crops in the natural environment could become agriculturally problematic. This can be prevented by thorough testing, first in the lab, then in a closed greenhouse followed by contained field testing and finally an extensive period of monitoring once the GM crops are being cultivated on a large scale. Except for the latter, this is largely what is required by many governments. Concerning GM crops, the risk of rampant growth in the natural environment is very small, and most of them will not survive anyway. For example, rape-seed plants, both transgenic and conventional, were cultivated in a field and studied by researchers for 10 years without harvesting. After 5 years there was not a single GM plant to be seen, and after 10 years there were just a few conventional ones. Crawley et al. (Crawley, Brown, Hails, Kohn, & Rees, 2001) state: "Four different crops (oilseed rape, potato, maize and sugar beet) were grown in 12 different habitats and monitored over a period of 10 years. In no case were the genetically modified plants found to be more invasive or more persistent than their conventional counterparts." Their results concern GM traits (resistance to herbicides or insects) that were not expected to increase plant fitness in natural habitats.

According to them this indicates that arable crops are unlikely to survive for long outside cultivations, but it does not mean that other genetic modification could not increase weediness or invasiveness of crop plants. They conclude: "The ecological impact of plants with GM traits such as drought tolerance or pest resistance that might be expected to enhance performance under field conditions will need to be assessed experimentally when such plants are developed." The same holds for GM plants for phytoremediation (Gressel & Al-Ahmad, 2005) and GM plants for production of pharmaceuticals. GM crops usually differ from their conventional counterparts only with respect to one or a few desirable genes, in contrast to crops from traditional breeding methods that mix thousands of genes (Atherton, 2002). Armed with genomic information and nanotechnology, plant molecular biologists are redesigning molecular toolkits to engineer plants still more precisely (Moeller & Wang, 2008). It would thus seem logical that GM crops pose fewer risks than conventionally modified crops.

Substantial scientific data on the environmental effects of the currently commercialised GM crops are available. Sanvido et al. (2007) have reviewed this scientific knowledge derived from the first 10 years of worldwide experimental field research and commercial cultivation. The review focuses on the currently commercially available GM crops that could be relevant for agriculture in Western and Central Europe (i.e. maize, rapeseed and soybean) and on the two main GM traits that are currently commercialised, i.e. herbicide tolerance and insect resistance. The sources of information include peer-reviewed scientific journals, scientific books, reports from regions with extensive GM-crop cultivation, as well as governmental reports. The data available so far provide no scientific evidence that the cultivation of the presently commercialised GM crops has caused environmental harm. The authors recognise, though, that results from large-scale cultivation systems, which GM crops generally are, have to be transferred with care to small-scale agricultural systems like in Switzerland. The interpretation of results is often challenged by the absence of a baseline for the comparison of environmental effects of GM crops in the context of modern agricultural systems (Sanvido et al., 2007). There is thus a need to develop scientific criteria for the evaluation of the effects of GM crops on the environment to assist the regulatory authorities. In their study, Sanvido et al. discuss the effects of GM-crop cultivation on the environment by considering the impacts caused by cultivation practices of modern agricultural systems. Even without GM crops, modern agricultural systems have profound impacts on all environmental resources, including negative impacts on biodiversity. When discussing the risks of GM crops, one has thus to recognise that the real choice for farmers and consumers is not between GM technology that may have risks and a completely safe alternative. The real choice is between GM crops and current conventional practices for pest and weed management, all possibly having positive and negative outcomes. To ensure a true precautionary policy, one should compare the risk of adopting a technology with the risk of not adopting it. This all led us to this fourth item requiring attention.

TEXTBOX 3.4.
The precautionary principle.

In November 2005 an interesting report on this subject was published by the Institute of Advanced Studies, United Nations University[19]. The report is called "Trading Precaution: The Precautionary Principle and the WTO". In the introduction it is stated that the precautionary principle is central to environmental policy and a key element in multilateral environmental treaties. As such it is a fundamental part of the Cartagena Protocol[20] on Biosafety. Policy makers and officials who use the precautionary principle and are involved in environmental and health matters assume

that precautionary measures must also be met if there is insufficient scientific proof of harm, but inaction may lead to irreversible damage or risk for health and environment. Conversely, the UN Millennium Task Force on Science, Technology and Innovation states in its report of 2005 that the focus on technological risks may overshadow the possible benefits of an up-and-coming technology, because these are often difficult to predict. Underlying the continuing debate on the precautionary principle is the fundamental question of how policy concerning health, safety and environment should be developed if on the one hand there is a lack of scientific consensus and on the other a significant portion of the population has irrational (from a scientific perspective) opinions

[19] www.ias.unu.edu
[20] bch.cbd.int/protocol

(fears) about (for) the material concerned. The precautionary principle endeavours to bridge the gap between scientific uncertainty and regulation of risk. Circumstances determine the way in which precautions are to be taken. These considerations make it difficult to draw up a generally applicable definition of the precautionary principle. International lawyers writing about the precautionary principle usually start from two ostensibly similar definitions. The first comes from the Bergen Ministerial Declaration on Sustainable Development of 1990:

"In order to achieve sustainable development, policies must be based on the precautionary principle. Environmental measures must anticipate, prevent, and attack the causes of environmental degradation. Where there are threats of serious or irreversible damage, lack of full scientific certainty should not be used as a reason for postponing measures to prevent environmental degradation."

The second oft-quoted definition is to be found in Principle 15 of the Rio Declaration on Environment and Development, of 1992:

"In order to protect the environment, the precautionary approach shall be widely applied by States according to their capabilities. Where there are threats of serious or irreversible damage, lack of full scientific certainty shall not be used as a reason for postponing cost-effective measures to prevent environmental degradation."

The major difference is in the word 'cost-effective', linking the need to take measures with the possible economic effect. The debate on the precautionary principle is complex and often abstract. In a certain sense the precautionary principle can be seen as a "rather shambolic concept ... muddled in policy advice and subject to whims of international diplomacy and the unpredictable public mood over the true cost of sustainable living" (O'Riordan & Cameron, 1994). In any case, the result was various different pieces of legislation between the EU and the US. As such the EU has very strict rules on authorisation and marketing of genetically modified organisms and products compared to the US. Conversely, some food products such as unpasteurised cheese are heavily regulated in the US for health reasons, while they are highly valued in the EU.

---

5. Development of a SMART legislative framework

Despite the ongoing controversies, in 2009 the area planted with GM crops grew still further. More than 13 million farmers in 25 countries planted GM crops, over 90% of them in developing nations (Marshall, 2009). This rapid progress is causing more anxiety about the effect on the environment, especially in Europe, where the number of field trials even fell and only a handful of the 27 EU countries cultivated the only GM crop approved there (Bt maize). In this context Richmond (2008) reviews the precautionary principle and believes that the progress made in every area of biotechnology quickly leads to countless applications and products to benefit the society. Progress is so rapid that policymakers, legislative authorities and law enforcers cannot keep up. This harbours the risk of serious and irreversible environmental consequences that will be difficult to control. The challenge is to develop a legislative framework with effective checks and balances that help avoid serious and possibly irreversible consequences but, at the same time, do not restrict innovation. The precautionary principle demands scientifically acceptable evidence that no damage will be done, if products are introduced or activities implemented. Determining scientific standards and norms (for example, less than 5% chance of damaging effects) by (i) asking the right questions and (ii) producing an acceptable experimental design, will lead to an approach that can reduce risk and provide policymakers and the public with a better understanding of possible problems in the future. Richmond also believes that, because there is a need to better understand and evaluate risks, statistics will be required as a framework for decision making (Textbox 3.5). In line with this we propose the SMART approach as the fifth item needing attention.

The SMART tool for defining goals[21] is well-known in human resource management. Analogously, it must be possible to base the legislation concerning agro-biotechnology on its own version of the SMART principle:

Sustainable–**M**easurable–**A**cceptable–**R**easonable–**T**ime-based

Many of the elements of the SMART approach already exist in the EU 2000 policy on the application of the precautionary principle. The EU Commission issued a policy communiqué in 2000 outlining "the Commission's approach to using the precautionary principle" and to "establish Commission guidelines for applying it" - Communication from the Commission on the Precautionary Principle COM 1. These guidelines

---

TEXTBOX 3.5.
Statistics as a framework for decision making.

There are two overall categories of statistical errors: the rejection of a correct hypothesis (Type I) and the acceptance of an incorrect hypothesis (Type II). For example, there is a gun on the table and there is no information available to establish whether or not it is loaded. The precautionary principle dictates that it must be assumed that all guns are loaded unless the opposite is proven. The alternative approach (often used for environmental considerations) is that everything is

safe until proven otherwise. If the gun was empty, but I have accepted the incorrect hypothesis that it was loaded, I am guilty of statistical error Type II. If the gun was indeed loaded and somebody suggested that it wasn't, that person was guilty of a Type I error, and, if the trigger is pulled, may also be guilty of murder. If we now replace the gun with open cultivation of transgenic crops, the doom scenario is clear. The lack of sufficient data to show that something is harmful doesn't mean that it is safe; the correct conclusion is that there are insufficient data to make a judgement.

[21] www.topachievement.com/smart.html

clearly state that the precautionary principle should be applied in a proportional, non-discriminatory and consistent manner, with an examination of the benefits and costs of action (or lack of action) and with an examination of scientific developments. It is interesting to note that the EU has failed to live up to its own policy. A good starting point for the development of a globally uniform, SMART-based legislative framework is the critical and thorough review by Chandler and Dunwell (2008) of hundreds of scientific papers on gene flow, risk assessment and environmental release of GM plants. A good model to start working with is wheat (Peterson & Shama, 2005). Wheat varieties produced with modern biotechnologies, such as genetic engineering and mutagenic techniques, have lagged behind other crop species and have only emerged recently. This offers a unique opportunity to assess comparatively the potential environmental risks (human health, ecological, and livestock risks) associated with genetically engineered, mutagenic, and conventional wheat production systems.

A problem hampering this development of the SMART approach, is the difference between the regulatory structures underlying US and EU policies regarding GM foods/crops. The US regulates GM foods/crops more as end products, applying roughly the same regulatory framework as to conventional ones. The EU, contrarily, regulates products of agro-biotechnology more as the result of a specific production process. Accordingly, EU regulates GM foods/crops specifically. As a result, the pertinent US regulation is relatively permissive, whereas EU regulation is relatively restrictive.

Both Ramjoué (2007) and Hammitt et al. (Hammitt, Wiener, Swedlow, Kall, & Zhou, 2005) analyse why GM food policies in the US and the EU are different. The fact is that the public debate in Europe has ground to a halt, having been reduced to a hopeless tug-of-war about GM foods/crops. A poignant example is the overwhelming majority voting in early 2009 against the proposals to overturn national bans on GM-maize cultivation in France, Greece, Austria and Hungary (Abbott, 2009). This EU impasse over agro-

EU MEMBERS IN TRENCH WARFARE OVER AGRO-BIOTECHNOLOGY

biotechnology was deepened even further in April 2009 with the ban on GM maize by the German government. In September 2009 Commission President José Manuel Barroso started an initiative to develop rules allowing member states to ban the cultivation of EU-approved crops. Proposals are due mid-2010. Despite of all this hassle, we as authors still strongly belief in employing agro-biotechnology and we challenge our colleagues to facilitate responsible progress and to inform the public objectively.

6. Responsible progress hand in hand with ongoing public debate.

The advocates and opponents of modern biotechnology need to stop fighting and start agreeing on SMART goals for the future. This means specifically and quantitatively defining which risks (if any) are acceptable, what is meant by "substantially equivalent" in the principle of substantial equivalence, a heavily criticised principle[22], what is of consequence in genetic modification of crops, what is sustainable, and what is natural or organic. To achieve this, the advocates and more especially the experts really need to understand that the public has both justifiable and imaginary concerns; this must be respected. In addition, the opponents should accept that GM plants are here to stay and can even offer huge benefits if we deal with these new technologies skilfully and carefully. Subsidised activist organisations such as Greenpeace are necessary as a counterbalance in a technological society such as the EU in general and the Netherlands

[22] www.i-sis.org.uk/subst.php

in particular, where they have become strong both in politics and communication. Furthermore, they have much more knowledge of these complex matters than they generally demonstrate in public. That is a hopeful sign, as is the fact that public opposition seems to be falling away. In the Eurobarometer public opinion survey of 2008, the percentage of those who said they were against GM crops fell from 70 to 58% (Abbott, 2009).

SEED COMPANIES HAVE THE POWER TO CONQUER THE WORLD

7. Integrated approaches for Third-World countries.

The domination of global agriculture by a handful of multinationals can have adverse effects, especially on small farmers in the Third World. Anti-trust laws should prevent this happening, but companies like Monsanto with the patents on the GM plants, have great power and could in a doom scenario take over the world in their grip via the food supply chain. During the last 30 years, we have seen that Third-World farmers are able to adopt new, more efficient technologies and really use them. However, it is still true that due to gene technology, agriculture has become even more dependent on a smaller number of large companies. We therefore feel justified in asking whether it is desirable for the situation to continue in this way. The first issue to address then is the plausibility of the claim that GM technology has the potential to provide the hungry with sufficient food for subsistence. Carter (2007) discusses this claim within the domain of moral philosophy to determine whether there exists a moral obligation to pursue this end if and only if the technology proves to be relatively safe and effective. By using Peter Singer's duty of moral rescue, she argues that we have a moral duty to assist the Third World through the distribution of GM plants. She concludes her paper by demonstrating that her argument can be supported by applying a version of the precautionary principle on the grounds that doing nothing might be worse for the current situation. Asante (2008) criticises opinions and perceptions blocking GM technologies that can potentially improve survival and quality of life for millions of people in Africa. We endorse his view that scientists must help provide an answer to this problem by ensuring that debate on GM crops addresses facts, not opinions. The initial refusal of badly needed food by some African countries in 2002 makes clear that most of them simply do not as yet have the experience and scientific capacity to make informed decisions about GM food. However, it is not only a lack

of experience with scientific decision making that makes Africa hesitant; some of the fears of the new science have their roots in mistakes in the past. Europeans, for instance, introduced water hyacinth and Nile perch in Africa with devastating consequences. So how can Africans be sure that GM foods/crops will not lead to even bigger mistakes? African governments can take a number of measures to prevent this, for instance by building a critical mass of people with the ability to evaluate and manage technology within the individual countries themselves. A strong scientific community will help select the best and most useful biotech applications and avoid any for which the risks outweigh the benefits. In the southern part of Africa alone where current food production is under the threat of climate change (Lobell et al., 2008), around 4 million people depend for their existence on food donations (Botha & Viljoen, 2008). Knowing this, it makes sense to consider GM food/crops as a means of reducing hunger and improving food quality. Africa did not profit from the Green Revolution that took place in the West in the middle of the last century. The expectation is that gene modification of traditional African food crops such as sorghum will produce a second green revolution from which they *will* benefit. The entire subject of GM organisms/technology is however saddled with different opinions, considerable frustrations, and growing ethical and environmental concerns, globally, leading to the already mentioned problem addressed by Asante (2008). Scientists in the individual African countries, and more particularly scientists from the West, must help to ensure that debates on GM crops address facts, not opinions.

SUPERDENSE SORGHUM...
OVERLOADED WITH VITAMINS!

After maize, wheat, rice and barley, sorghum is the most important grain in the world and the second most important crop on the African continent. In 2006 the Bill and Melinda Gates Foundation donated $450 million to the African Biotechnology Sorghum consortium that consists of companies, institutes and universities in Southern Africa and North America (Botha & Viljoen, 2008). It aims at using gene technology to improve the health and welfare of people in the poorest countries of the world by making GM sorghum that is more nutritional and more digestible. The target is a GM sorghum that contains more essential amino acids, especially lysine, but also increased levels of vitamin A and E, and more absorbable forms of iron and zinc.

Botha and Viljoen (2008) have carefully analysed all the advantages and disadvantages, making use of the experience gained with Golden Rice. They conclude

that it is doubtful whether the development costs of this GM sorghum can be justified when compared with the costs of investing in sustainable African agriculture. According to them, GM sorghum can only be successfully introduced if it forms part of an integrated approach. The Alliance for Green Revolution in Africa shares this vision. In less than two years, according to Kofi Annan, this organisation has collected $330 million for a comprehensive and integrated project, initially in the following six areas[23]:

1. Development of higher yielding, disease-resistant and climate-resilient varieties of African crops.
2. Seed-multiplication and distribution systems.
3. Improved soil health.
4. Agricultural education.
5. Agro-dealer networks that get inputs to farmers in remote locations.
6. Development of policies that benefit small-hold farmers.

Issues that include water use, food storage and processing, and market development are also considered.

As in the rest of the world, examples of GM food that are beneficial for the health of individual consumers are badly needed in Africa and other Third-World countries. Biofortified sorghum is a good start. Naqvi et al. (2009) reported recently on orange maize with extra vitamins. Using gene technology German and Spanish researchers have enriched South African white maize

[23] www.nrc.nl/redactie/binnenland/speeches/kofi_annan.pdf

with beta carotene (a precursor of vitamin A causing the orange colour of the maize), and precursors of vitamin C and folic acid. Natural white maize, which is a staple food in many developing countries, contains relatively few vitamins. Compared to the normal maize, this GM maize contains as precursor equivalents six times as much vitamin C, twice as much folic acid, and 169 times as much vitamin A. This means that consumption of 100 to 200 grams of the GM maize yields the daily recommended amount of vitamin A and folic acid and 20% of that of vitamin C.

3.3. CONCLUSIONS

In April 2009 a Dutch proposal concerning whether or not the decision to cultivate GM crops should be left to individual Member States (Anonymous, 2009), was put forward to the EU Council. The then Czech presidency said that a surprising number of countries reacted positively to it. The proposal suggested that a possible solution to GM crops approval issues would be to apply internal market rules for the import of products – with a decision on the EU level, but for cultivation it could be left to each Member State. In September 2009, after the elections, the new Commission President José Manuel Barroso started an initiative to indeed develop rules allowing the separate member states to ban the cultivation of EU approved crops. The appointment of John Dalli as Commissioner for Health and Consumer policy clearly showed a shift from an anti-to pro-GM crop policy. Less than a month in office he had already taken the most controversial decision a euro-commissioner can take: at the beginning of 2010 he approved the cultivation of a second GM crop, i.e. the Amflora potato of BASF. For twelve years all decisions on approvals were halted. In mid-July 2010, at the time of finishing this chapter, he came with a new law proposal giving the separate countries authority to ban GM crops. According to experts, this proposal creates political room to approve GM crops faster at the EU level. It gives us the feeling anyway, seeing this all happen, that the Member States are moving slowly towards a consensus on lifting the bans, which is indispensable for responsible progress at least in some of the Member States. The point at which a firm "yes" will be obtained from all members still seems a long way off, but we believe that it is still not too late, if we pay sufficient attention to the seven points elaborated in this chapter.

3.4. SOURCES

Abbott, A. (2009). European disarray on transgenic crops. *Nature, 457*(7232), 946-947.

Abhilash, P. C., Jamil, S., & Singh, N. (2009). Transgenic plants for enhanced biodegradation and phytoremediation of organic xenobiotics. *Biotechnology Advances, 27*(4), 474-488.

Anon. (2009). Member States moving to consensus on bans. *AgBiotech Reporter, 2*(7), 1,3-4.

Asante, D. K. A. (2008). Genetically modified food - The dilemma of Africa. *African Journal of Biotechnology, 7*(9), 1204-1211.

Atherton, K. T. (2002). Safety assessment of genetically modified crops. *Toxicology, 181*, 421-426.

Batista, R., & Oliveira, M. M. (2009). Facts and fiction of genetically engineered food. *Trends in Biotechnology, 27*(5), 277-286.

Boddiger, D. (2007). Boosting biofluel crops could threaten food security. *Lancet, 370*(9591), 923-924.

Botha, G. M., & Viljoen, C. D. (2008). Can GM sorghum impact Africa? *Trends in Biotechnology, 26*(2), 64-69.

Butelli, E., Titta, L., Giorgio, M., Mock, H. P., Matros, A., Peterek, S., Schijlen, E. G. W. M., Hall, R. D., Bovy, A. G., Luo, J., & Martin, C. (2008). Enrichment of tomato fruit with health-promoting anthocyanins by expression of select transcription factors. *Nature Biotechnology, 26*(11), 1301-1308.

Carter, L. (2007). A case for a duty to feed the hungry: GM plants and the third world. *Science and Engineering Ethics, 13*(1), 69-82.

Chandler, S., & Dunwell, J. M. (2008). Gene flow, risk assessment and the environmental release of transgenic plants. *Critical Reviews in Plant Sciences, 27*(1), 25-49.

Crawley, M. J., Brown, S. L., Hails, R. S., Kohn, D. D., & Rees, M. (2001). Biotechnology - Transgenic crops in natural habitats. *Nature, 409*(6821), 682-683.

Domingo, J. L. (2007). Toxicity studies of genetically modified plants: A review of the published literature. *Critical Reviews in Food Science and Nutrition, 47*, 721-733.

Dona, A., & Arvanitoyannis, I. S. (2009). Health Risks of Genetically Modified Foods. *Critical Reviews in Food Science and Nutrition, 49*(2), 164-175.

Fedoroff, N. (2008). Seeds of a perfect storm. *Science, 320*(5875), 425-425.

Fox, J. L. (2004). Vatican debates agbiotech. *Nature Biotechnology, 22*(1), 4-5.

Gressel, J. (2009). Is FAO selling biotech short on biofuels? *Nature Biotechnology, 27*(1), 22-23.

Gressel, J., & Al-Ahmad, H. (2005). Assessing and managing biological risks of plants used for bioremediation, including risks of transgene flow. *Zeitschrift für Naturforschung Section C - A. Journal of Biosciences, 60*, 154-165.

Hammitt, J. K., Wiener, J. B., Swedlow, B., Kall, D., & Zhou, Z. (2005). Precautionary regulation in Europe and the United States: A quantitative comparison. *Risk Analysis, 25*(5), 1215-1228.

James, C., & Strand, S. (2009). Phytoremediation of small organic contaminants using transgenic plants. *Current Opinion in Biotechnology, 20*(2), 237-241.

Kok, E. J., Keijer, J., Kleter, G. A., & Kuiper, H. A. (2008). Comparative safety assessment of plant-derived foods. *Regulatory Toxicology and Pharmacology, 50*(1), 98-113.

Konig, A., Cockburn, A., Crevel, R. W. R., Debruyne, E., Grafstroem, R., Hammerling, U., Kimber, I., Knudsen, I., Kuiper, H. A., Peijnenburg, A., Penninks, A. H., Poulsen, M., Schauzu, M., & Wal, J. M. (2004). Assessment of the safety of foods derived from genetically modified (GM) crops. *Food and Chemical Toxicology, 42*, 1047-1088.

Kuiper, H. A., Kok, E. J., & Engel, K. H. (2003). Exploitation of molecular profiling techniques for GM food safety assessment. *Current Opinion in Biotechnology, 14*(2), 238-243.

Lobell, D. B., Burke, M. B., Tebaldi, C., Mastrandrea, M. D., Falcon, W. P., & Naylor, R. L. (2008). Prioritizing climate change adaptation needs for food security in 2030. *Science, 319*(5863), 607-610.

Macek, T., Kotrba, P., Svatos, A., Novakova, M., Demnerova, K., & Mackova, M. (2008). Novel roles for genetically modified plants in environmental protection. *Trends in Biotechnology, 26*(3), 146-152.

Marshall, A. (2009). 13.3 million farmers cultivate GM crops. *Nat Biotechnology, 27*(3), 221-221.

Meldolesi, A. (2009). Vatican cheers GM. *Nature Biotechnology, 27*(3), 214-214.

Miller, H. I. (2008). Auf Wiedersehen, agbiotech. *Nature Biotechnology, 26*(9), 974-975.

Miller, H. I. (2009). A golden opportunity, squandered. *Trends in Biotechnology, 27*(3), 129-130.

Miller, H. I., Morandini, P., & Ammann, K. (2008). Is biotechnology a victim of anti-science bias in scientific journals? *Trends in Biotechnology, 26*(3), 122-125.

Miller, H. I., Seaton, B. A., Carlile, S., Kaiserlian, D., Bankaitis, V., Corbi, A., Ezekowitz, R. A. B., Fridman, W. H., Funnell, B., & Gettins, P. G. W. (1997). *Policy controversy in biotechnology: An insider's view*, Academic Press.

Moeller, L., & Wang, K. (2008). Engineering with precision: Tools for the new generation of transgenic crops. *Bioscience, 58*(5), 391-401.

Naqvi, S., Zhu, C., Farre, G., Ramessar, K., Bassie, L., Breitenbach, J., Perez Conesa, D., Ros, G., Sandmann, G., Capell, T., & Christou, P. (2009). Transgenic multivitamin corn through biofortification of endosperm with three vitamins representing three distinct metabolic pathways. *Proceedings of the National Academy of Sciences of the United States of America, 106*, 7762-7767.

O'Riordan, T., & Cameron, J. (1994). *Interpreting the precautionary principle*. London, Earthscan/James & James.

Peterson, R. K. D., & Shama, L. M. (2005). A comparative risk assessment of genetically engineered, mutagenic, and conventional wheat production systems. *Transgenic Research, 14*(6), 859-875.

Ramjoue, C. (2007). The transatlantic rift in genetically modified food policy. *Journal of Agricultural and Environmental Ethics, 20*, 419-436.

Richmond, R. H. (2008). Environmental protection: applying the precautionary principle and proactive regulation to biotechnology. *Trends in Biotechnology, 26*(8), 460-467.

Sanvido, O., Romeis, J., & Bigler, F. (2007). Ecological impacts of genetically modified crops: ten years of field research and commercial cultivation. *Advances in Biochemical Engineering/Biotechnology, 107*, 235-278.

Strauss, S. H., Tan, H., Boerjan, W., & Sedjo, R. (2009). Strangled at birth? Forest biotech and the Convention on Biological Diversity. *Nature Biotechnology, 27*(6), 519-527.

Walter, C., Fladung, M., & Boerjan, W. (2010). The 20-year environmental safety record of GM trees. *Nature Biotechnology, 28*(7), 656-658.

part two
Our daily food and drink

"There has never been any suggestion that genetically manipulated food is harmful to the consumer. And yet there are still serious concerns about it. Europe now needs to determine whether the truth is closer to the gloomy pronouncements of Greenpeace or the risk-free Teletubby-like utopia that the biotech industry presents."

Rik Nijland, science writer, April 1999

Modern biotechnology is clearly a very hot topic in this present day and age, particularly where our daily food and drink are concerned. Traditional biotechnology has played an important role in our food production for centuries, but modern biotechnology has now become an unavoidable part of this process. However, the heated discussions are chiefly concerned with food from transgenic crops, the so-called gene food, variously called *Franken(stein) Food* or monster food by its opponents. In part two of the book gene food is dealt with separately in Chapter 8, as are the traditional biotechnological products like cheese, bread, wine and meat. For here too, modern biotechnology now plays a key role.

"As their highnesses travelled", wrote Horace Walpole in an 18th century letter to a friend, commenting on a fairy tale he had been reading, "they were always making discoveries, by accidents or sagacity, of things they were not in quest of."

It was Walpole who suggested that the word "serendipity" be included in our vocabulary after reading the Three Princes of Serendip. Serendip is the old Persian name for Sri Lanka. Nowadays serendipity is defined as the finding of something unexpected and useful particularly whilst looking for something entirely unrelated, or to use the visual words of Pek van Andel, studying serendipity and a winner of the Ig Nobel prize: "looking for a needle in a haystack and rolling out of it with a milkmaid." Since 1994, Serendip has also been an interactive educational website that helps people improve their chances of deliberately making discoveries by chance[24]. The discovery of cheese is a notable example of serendipity.

24 serendip.brynmawr.edu/serendip

4.1. OLD CHEESE

Biotechnology is at least as old as documented history. Before 700 BC Homer, the author of The Iliad and The Odyssey, the oldest preserved examples of Greek literature, described a simple, yet interesting biotechnological experiment. He wrote that if you crush a fig branch and then stir the crushed part into milk, a solid forms in the milk, leaving a fluid which can then be drained off. What he was describing here is the making of a type of cottage cheese. What Homer didn't and couldn't know, is that the crushed fig branch oozed a little sap which contained the enzyme ficain (or ficin). This enzyme causes the casein (curds), the components in milk that help form cheese, to separate for the most part, thus making the casein curdle.

MAKING CHEESE BY USING CRUSHED FIG BRANCHES IS A VERY OLD TECHNIQUE....

EVE, ARE YOU MAKING CHEESE AGAIN?!

From an even earlier age comes yet another cheese story. By removing the fourth stomach of a freshly slaughtered young calf and pouring milk into it, the same process can be observed: the casein in the milk curdles. Here too a similar kind of enzyme is responsible for this action, namely chymosin (also called rennin), which leaks from the stomach wall and enters the milk. The chymosin then divides up the casein into a large part (90%) that separates out and a small part (10%) that remains dissolved in the residual liquid (whey). This must have been a mysterious but useful occurrence for observers in ancient times. As far as we're concerned it is one of the first ever biotechnological applications.

4.2. TRADITIONAL CURDLING

In the nineteenth century there emerged a little understanding of what curdling actually involved. Furthermore, the first curdling company was founded in that century, in 1875 in Copenhagen by a man called Christian Hansen. Hansen bought rennet stomachs from freshly slaughtered young calves and, using salt solution, extracted the chymosin from them. The extract, rennet, is one of the first standardised, industrial products to be used in a biotechnological process, i.e. cheese-making. The Christian Hansen company is still producing rennet today, in virtually the same way. However, since 2002, the company has been working on new developments in collaboration with Novozymes, also a Danish company and one of the world's biggest enzyme manufacturers, which makes frequent use of modern biotechnology (Textbox 4.1).

TEXTBOX 4.1.

Cheese alliance.[25]

On 26 August 2002 Novozymes announced an alliance with the international food ingredients company Chr. Hansen, with the initial aim of boosting yields in cheese production. The first fruit of this collaboration was launched in 2005. The product in question was a phospholipase (hydrolysing enzyme) which was brought onto the market under the name of YieldMAX PL. The enzyme is added to milk as a pre-treatment process to optimise coagulation and give a higher yield of cheese. The yield increase is in the order of two percent. That may seem trifling to the layman, but cheese professionals regard it as one of the greatest innovations for several decades, given that all other attempts from the whole dairy industry in the previous ten years have only delivered a one percent increase in yield. The enzyme works especially well in cheeses like mozzarella (BioTimes December 2005; newsletter published by Novozymes).

For every 10,000 litres of milk, approximately 1 litre of rennet is used in cheese-making. That may not sound like much, but considering that in the Netherlands alone 700,000 tonnes of cheese are produced each year, starting with 7,000,000,000 litres of milk, for which 700,000 litres of rennet are needed (approx. 1 litre of rennet per tonne of cheese), then it soon becomes clear just how many calves' stomachs are required. Rennet is therefore a scarce and expensive commodity and the industry has long been anxiously searching for an alternative. With the single exception of the enzyme from the microorganism *Mucor miehei,* attempts to bring microbial rennet onto the market had all been fairly unsuccessful. Until twenty years ago in 1989, that is, when a Dutch company (the present-day DSM-Gist), brought an alternative rennet onto the market which was in quality terms at least as good as the natural version. The basis of the technique used to make this new product was laid down in 1973, once the first successful recombinant DNA experiments had been conducted. Ironically, Dutch cheese-makers are among the few in the industry who still don't use this technique (find out why in the next section). This is even more remarkable because many of the enzymes used in food production are currently made by using recombinant microorganisms (Olempska-Beer, Merker, Ditto, & DiNovi, 2006). We will come back to them in the next chapter about bread.

4.3. MODERN CURDLING

In the early 1980's the Dutch biotechnology company, Gist-Brocades (now part of DSM), began experimenting with recombinant DNA technology. Researchers at Gist-Brocades bought from Unilever the chymosin gene of a cow; this gene is the piece of DNA that ensures the production of the enzyme chymosin in suckling calves. They then "inserted" this piece into the genetic material of yeast cells from *Kluyveromyces lactis,* one of their so-called "plugbugs" (Textbox 4.2). These

[25] www.novozymes.com/NR/exeres/11CACD69-CDAE-4959-94F9-CECD11D83C66.htm

plugbugs then made calf chymosin. The daughter cells of these yeast cells, which can be cultivated in great numbers in huge fermentation vats, also produce calf chymosin. These genetically modified yeast cells have been used for the past two decades to produce very pure rennet enzyme, which is identical to the authentic calf chymosin. Extensive testing has shown that this product of modern biotechnology can be used safely with no risk to health and that it works at least as well as the traditional rennin.

Switzerland was the first country to use this new chymosin. That was in the late '80s, when the public was probably still largely unaware of what sort of product it was. Now the people of this country are very sceptical about modern biotechnology. Meanwhile, this product has been accepted and is used in many countries around the world, while other companies have also come onto the market with bovine chymosin, made with recombinant microorganisms. France is one of the countries that held back approval for a long time, but then gave permission in 1998 following the "Mad Cow Disease" episode (the risk of such infection via rennin cannot be excluded).

Remarkably enough, the production of chymosin from genetically modified microorganisms has stimulated cheese consumption in Israel and the United States. The microbial product has been declared *kosher*. Religious Jews can, and therefore do, eat this cheese. Muslims may also eat cheese made with this chymosin, because it is also *halal* (i.e. meets the Islamic criteria concerning food preparation). Even Professor Lucas Reijnders of the Dutch Institute for Nature and the Environment declared back in 1994 that he was in favour of the use of the recombinant enzyme, because it meant that vegetarians could eat cheese made from it without fear of betraying their principles.

Ironically, the Netherlands is one of the few countries in which cheese-makers still don't use it. Although incredibly late in the day for a country where it was first produced and where cheese production is among the highest in the world, permission for its use was finally granted in 1992. That said, our domestic cheese producers still don't use it, fearing that the German consumers will stop buying our cheese. Germany is one of the Netherlands' biggest customers, but also the country where opposition to anything involving modern biotechnology has been very pronounced since the beginning of the lobby against gene technology. Despite that, the use of recombinant chymosin has also been permitted in Germany since 1997. Although cheese manufacturers in both countries are very reluctant to change, consumers have been buying cheeses made with the recombinant enzyme for years, as many cheese manufacturers in other countries use it and these cheeses are currently very popular among many consumers all over the world. However, the percentage of end-users who are actually aware that modern biotechnology has been used to make these cheeses, is small. The gentleman's agreement between the Dutch dairy companies not to use the recombinant enzyme seems now, though, to be on shaky ground, because in recent times recombinant chymosin has been used on a small scale in Germany. That is probably why its use in the Netherlands will not be long in coming.

The recombinant microorganisms which DSM-Gist works with primarily, come from normal microbial strains which this company has been using for years on a commercial scale to produce enzymes, and thus has plenty of experience and knowledge of it. These microorganisms, or "bugs" as they are popularly called, have a number of advantages which they have been selected on over the years. These include efficient secretion of enzymes and the certainty that they are classified as safe organisms. They are GRAS: "Generally Recognized As Safe". The recombinant microorganisms derived from them can be used in the same fermentors and reprocessing and purification apparatus as the non-recombinant strains, on the understanding that they take place under restrictive conditions, i.e. depending on the process, under more or less stringent safety requirements, as prescribed by the law on the use of recombinant organisms. An added advantage of these bugs is that they have been modified so that they have special food requirements, which means that in the unlikely event of them getting out of the fermentor, they will not be able to continue to grow outside. DSM-Gist has given this technology the trade name 'PluGbug®', reflecting the ease with which extra genes can be plugged into these bugs[26].

BUILD YOUR OWN MICROORGANISM.

LEGO PLUGBUG

GRAS

The yeast Kluyveromyces lactis is one of DSM-Gist's plugbugs. As we saw above, a recombinant form is used for making curdling enzyme. This yeast has been used for more than 30 years to produce the enzyme lactase, which is used to convert lactose in milk products into galactose and glucose. These are sugars that can be better digested by people who have lactose intolerance (major sections of the population in Asia and Africa are lactose-intolerant), so that these people can also consume milk products without suffering side effects. At the beginning of the 1980s researchers at the company isolated not only the lactase gene, but also the DNA which is necessary to express the gene, and therefore the lactase. From this they constructed the so-called gene expression cassettes, on which they can easily record a gene of choice. This cassette can then be accurately inserted into the genome of the yeast, ensuring with almost complete certainty the expression of this new gene. They also performed something similar with the bacterium Bacillus licheniformis and the fungus Aspergillus niger. This opened the way to the efficient production of a whole range of proteins "foreign" to these plugbugs, for example enzymes like chymosin, and pharmaceuticals, etc. This information about the plugbug concept is taken from the Gist-brocades 1991 brochure 'Biotechnology, today and tomorrow', but more recent publications on this have appeared since, for example, Groot et al. (Groot, Herweijer, Simonetti, Selten, & Misset, 2000).

N.B. In the February 2007 issue of Nature Biotechnology

[26] www.dsm.com/en_US/html/dfs/genomics_at_dsm.htm

(Cullen, 2007) academic and DSM researchers published the complete sequence of the Aspergillus niger genome. This genome project also produced new application possibilities, namely the production of an enzyme that promotes muscle regrowth after sporting exertions and an enzyme that prevents the formation of the toxic substance acrylamide in baked and fried products (see Textbox 5.3).

4.4. CHEESE RIPENING: NOW AND IN THE FUTURE

When preparing cheese, as much protein and fat as possible must be separated out of the milk, and as soon as possible this must be set aside to ripen into cheese. Ripening is the complex process required for the development of a cheese's flavour, texture and aroma. Proteolysis, lipolysis and glycolysis are the main biochemical reactions that are responsible for the basic changes during the maturation period. As ripening is a relatively expensive process for the cheese industry, reducing maturation time without destroying the quality of the ripened cheese has economic and technological benefits. A review of traditional and modern methods used to accelerate Cheddar cheese ripening is presented by Azarnia *et al.* (2006).

As we have already established, the separation is activated by adding a milk-curdling enzyme, either from a genetically modified microorganism or in the form of a naturally-occurring enzyme. Early in the cheese-making process starter cultures (Textbox 4.3) are added along with rennet. The cultures are a mixture of lactic acid bacteria, whose composition varies from one cheese to another. Lactic acid bacteria play an important role in cheese ripening. Enzymes cleave the proteins into short pieces, peptides, which are then divided further into the individual amino acids (about 20 different ones in total form the building blocks of all natural proteins). These amino acids give the basic taste to the cheese, but can also later be converted into volatile (sulphurous) components with a strong cheese or cabbage taste.

As already said cheese maturation is a relatively slow process, because the enzymes required are only released when the lactic acid bacteria die and then break open, or lyse. The ripening process is thus expensive: the storage of cheese in conditioned areas costs the Netherlands alone more than ten million euros per week! It is hardly surprising then that the race is on to find new means of speeding up the process of cheese ripening. Elevated ripening temperatures, addition of enzymes, addition of cheese slurry, adjunct cultures, genetically engineered starters and recombinant enzymes and microencapsulation of ripening enzymes are traditional and modern approaches to accelerating cheese ripening (Azarnia *et al.*, 2006).

An approach used by DSM Food Specialties involves adding extra enzymes. In 1996 the company applied for a patent on the phenylalanine-aminopeptidase enzyme, produced by a non-genetically modified mould. This enzyme cleaves the amino acid phenylalanine from peptides; phenylalanine is an

Cheese ripening is catalysed by milk enzymes, coagulant, starter lactic acid bacteria and non-starter lactic acid bacteria. All milk components remaining in the curd are involved in the ripening, which involves the enzymatic degradation of these components. In general, the important components in cheese ripening are: chymosin or rennet substitutes, natural milk enzymes, starter bacteria and their enzymes, and enzymes from secondary starter cultures and moulds. Starter bacteria have an important role in the development of flavour. Because of their main role in the progressive acidification of cheese, increasing the number of starter bacteria can result in over-acidification of the final curd. Attenuated starter cultures are used for the purpose of reducing the acid-producing ability of the cells without the destruction of their intracellular enzymes.

---

amino acid that contributes to the taste of the cheese. The addition of phenylalanine-aminopeptidase to cheese milk shortens the maturation time of Emmental and Cheddar. From a marketing perspective, however, there is a problem. If the recipe is changed, the traditional cheese names can no longer be used. So DSM is focusing on the American market of enzyme-modified cheeses, which are made by grating young cheese and heating it up in the presence of taste-forming enzymes. These enzyme-modified cheeses can be perfectly processed into ingredients or flavourings for products like hamburgers and pizzas. On the DSM website[27] one can find their present starter cultures and dairy enzymes.

In a second approach to accelerating cheese ripening, genetically modified lactic acid bacteria are added to the starter culture. These bacteria can lyse 'to order' and then give up their enzymes to the cheese. This order can, for example, be given by the substance nisin (a preservative used in cheese preparation)

(Textbox 4.4), or an increase in the salt concentration or temperature. Theoretically, the use of these fast-lysing bacteria can shorten the maturation period of, for example, Gouda cheese by 75%. The use of this technique is, however, currently blocked because of the previously mentioned protection of type indications such as Gouda, but also because the chance of food containing genetically modified organisms (GMOs) being accepted by the consumer is still fairly small (Textbox 4.5).

A third approach being worked on is the addition to the starter cultures of lactic acid bacteria that overproduce certain enzymes. One such enzyme is cystathionine-β-lyase. This converts the sulphurous amino acid methionine into methanethiol, a direct precursor of volatile aromas in Gouda cheese. Both approaches with genetically modified lactic bacteria have given spectacular study results, but to our knowledge neither are being used yet. The future will decide if and when the general public will fully accept these "classic" products of modern biotechnology.

[27] www.dsm.com/en_US/html/dfsd/home.htm

Nisin is an antibacterial peptide that is produced naturally by some lactic acid bacteria in order to counteract the development of competing microorganisms. The action of nisin relies on the creation of permeable bacterial membranes (Figure 4.1). Since nisin is active against perishable and pathogenic microorganisms such as Clostridia and Listeria, it is used as a natural preservative, for example in cream cheese and in Eastern Europe in fruit and vegetable preserves. Once a lactic acid bacterium has formed nisin, this peptide appears to stimulate its own production. Nisin induces a membrane-bound sensor protein, NisK, to activate a regulator protein, NisR. This occurs via the transfer of a phosphorus group (Pi). The activated NisR then binds to the nisin promotor, P*. A promotor is a piece of the DNA in front of a gene or genes that regulates the action of this gene or genes. Normally the nisin gene is located behind this nisin promotor, so that

extra quantities of this peptide are made. The nisin gene can, however, be replaced by an arbitrary gene X via genetic modification. This is how the patented NIsin Controlled Expression or NICE system came about. In this system, the expression of gene X and the production of the accompanying protein X can be accurately controlled by the addition of more or less nisin (Zhou, Li, Ma, & Pan, 2006).

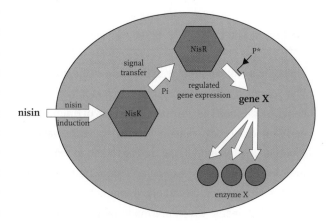

Figure 4.1. NICE gives control over the production of desired proteins such as enzymes.

4.5. THE FINAL QUESTION

In Chapter 2 we mentioned a development in New Zealand that provoked a strong reaction, namely the campaign by MAdGE (Mothers Against Genetic Engineering in Food and the Environment). The February 2003 issue of *Nature Biotechnology* reveals the scientific background. In this journal Karatzas from New Zealand published an article (Karatzas, 2003) stating that they produced nine transgenic cloned cows

with more casein protein in their milk. This extra protein in the milk means that more cheese can be produced more cheaply. As a result of this article, several Members of the Dutch House of Representatives put questions to the former Minister of Agriculture, such as whether he thought the genetic modification of animals for food production was ethical, whether there should be an overall testing framework to weigh up the pros and cons of this sort of development, and whether he could prevent these cows or their products from

TEXTBOX 4.5.

Acceptability of genetically modified cheese.

Many European consumers still have rather negative attitudes towards the use of gene technology in food production. In 2002 Scandinavian researchers published the study "Acceptability of genetically modified (gm) cheese as real product alternative" (Lahteenmaki et al., 2002). The objective of this study was to examine whether taste and health benefits influence the acceptability of genetically modified products when they are presented as real product alternatives. Consumers in Denmark, Finland, Norway and Sweden (n=738) assessed two cheeses: one was labelled as genetically modified (preferred in an earlier product test) and the other as conventional (neutral in an earlier product test). A smaller control group received two cheeses with blind codes. Labelling decreased consumers' intentions to buy the originally preferred GM-labelled cheese, but still the intentions were at the same level as the conventionally labelled option. Participants chose two GM cheeses out of five possible when given the option to take cheese home after tasting. Intentions to buy GM cheese could best be explained by respondents' attitudes towards gene technology and perceived taste benefits. General health interest was also a reinforcer of intentions for GM cheese with reduced fat content.

being imported into the Netherlands. In Chapter 2 we saw the emergence of precisely that sort of EU testing committee, namely the EFSA. As far as Dutch legislation is concerned, genetically modifying and cloning cows to improve cheese production is definitely not important enough, but the final question remains as to whether or not Dutch legislation can prevent the resulting products from being imported, given that the WTO treaties allow free trade when there are no scientific reasons, for instance with respect to safety, to ban them.

4.6. SOURCES

Azarnia, S., Robert, N., & Lee, B. (2006). Biotechnological methods to accelerate cheddar cheese ripening. *Critical Reviews in Biotechnology, 26*(3), 121-143.

Cullen, D. (2007). The genome of an industrial workhorse. *Nature Biotechnology, 25*(2), 189-190.

Groot, G. S. P., Herweijer, M. A., Simonetti, A. L. M., Selten, G. C. M., & Misset, O. (2000). Enzymes in food and feed: past, present and future. In Stanislaw Bielecki, Johannes Tramper & J. Polak (Eds.), *Progress in Biotechnology* (Vol. 17, pp. 95-99), Elsevier.

Karatzas, C. (2003). Designer milk from transgenic clones. *Nature Biotechnology, 21*(2), 138-139.

Lahteenmaki, L., Grunert, K., Ueland, O., Astrom, A., Arvola, A., & Bech-Larsen, T. (2002). Acceptability of genetically modified cheese presented as real product alternative. *Food Quality and Preference, 13*(7-8), 523-533.

Olempska-Beer, Z., Merker, R., Ditto, M., & DiNovi, M. (2006). Food-processing enzymes from recombinant microorganisms--a review. *Regulatory Toxicology and Pharmacology, 45*(2), 144-158.

Zhou, X., Li, W., Ma, G., & Pan, Y. (2006). The nisin-controlled gene expression system: construction, application and improvements. *Biotechnology Advances, 24*(3), 285-295.

BIOTECHNOLOGY IN THE BAKERY: ON THE RISE!

"You can only really say that something is safe if you yourself are convinced. And we are. The enzymes are not being tinkered with. And if the enzyme producers are doing that, we'll know about it."

This bold statement was issued by Esther Delnoij on 7 May 1994. At that time she was head R&D of a manufacturer of bread improvers. Like cheese, bread is one of the oldest traditional biotechnological products. In the last decade of the last century, however, modern biotechnology has also entered the baking industry in the form of recombinant enzymes as bread improvers and raw materials that may originate from genetically modified crops.

Our daily bread basically consists of flour, water, yeast and a little salt. No-one knows exactly when and how yeast was first put into bread to make it rise. It almost certainly happened by chance the first time, probably in the Nile valley at the time of the Pharaohs. What we do know is that the later Egyptians ate leavened bread and that the Old Testament is also clear on this subject. Here there is a description of bread with or without sourdough (with or without added yeast) (Textbox 5.1). As anyone who has ever tried to bake bread knows, bread leavened with yeast is different mainly in terms of the texture and structure, but also has a better aroma and taste. When yeast - a living, single-celled organism - is added to the dough mixture, the yeast cells grow, divide and thus increase in number. As the yeast grows, the cells ferment the sugars in the dough, producing carbon dioxide among other things. These gas bubbles are trapped in the dough thus forming the light texture of well-leavened bread that we know and love.

BREAD WITH YEAST WAS 'INVENTED' BY ACCIDENT IN ANCIENT EGYPT

MMM, LET'S SEE WHAT HAPPENS WHEN I PUT A SINGLE-CELLED ORGANISM INTO THIS DOUGH...

TEXTBOX 5.1.
Sourdough.

Wheat contains by nature different types of so-called wild yeast cells. However, the concentration of these yeast cells is so low that it is impossible to get a dough to rise with it. You can however let the concentration increase naturally. All you need to do to set this process in motion is add water to flour and ensure that there is enough oxygen in it. Due to the acetic acid and lactic acid bacteria that also occur naturally in flour, the acidity of the mixture increases, hence the name sourdough. After a few days you have grown a sourdough whose concentration of yeast cells is sufficient for the purposes of baking bread. To this end, the sourdough must be mixed with more flour and the mixture then left to stand for a day. The number of yeast cells is however never as high as in the baker's yeast that you can buy in the shops. That is a so-called pure culture - one type of yeast grown in large vats (fermentors) in a factory. One gram of this contains about ten billion yeast cells, while one kilo of flour only contains about 30 thousand yeast cells. Sourdough bread is therefore usually less light.

5.2. BAKER'S YEAST

Baker's yeast can rightly be regarded as one of the oldest products of industrial fermentation. The industrial-scale production of baker's yeast and its widespread use probably started with the Viennese process developed by Ignaz Mautner around 1846. To date approximately 500 different yeast types have been identified. As a result of its ability to produce large quantities of carbon dioxide, *Saccharomyces cerevisiae* is the most commonly used type in bakeries and for that reason is known as baker's yeast (Textbox 5.2). As far as volume and function are concerned it is one of the most important biotechnological products of all time. Every year more than two million tonnes of baker's yeast are produced around the world. Most of the yeast is used to make a large variety of bread types. It is also used for pastries, biscuits, crackers and pizzas.

Source: Getty Images

5.3. DOUGH

Wheat dough consists mainly of gluten (a protein network composed of gliadin and glutenin), lipids (fats), starch and other non-starch carbohydrates. This natural raw ingredient can vary tremendously in quality and also undergoes a great many process steps during bread preparation. Dough is developed as a result of various different processes. First of all, the kneading process breaks up the structure of the protein complex, which is formed after flour and water are mixed. The kneading stretches the protein chains and lines them up next to each other. During the rising process of the dough, they form a big protein network, called gluten. It is important that the gluten proteins are mixed well, as this determines the gas-holding capacity of the dough as well as its final volume and firmness. For centuries, the variation in the raw ingredients and the considerable number of processing steps have

TEXTBOX 5.2.
Baker's yeast.

This photo is an electron microscopic image of baker's yeast (Saccharomyces cerevisiae). Yeasts are unicellular fungi (Ascomycetes), which can survive in aerobic as well as anaerobic (without oxygen) conditions. They are important for breweries and bakeries because of the alcohol and carbon dioxide that they produce as a result of respiration. Reproduction is normally asexual by means of budding; the buds are clearly visible on the cells in the photo.

made it difficult for bakers to bake consistently good quality bread.

CONSISTENT BREAD QUALITY USED TO BE VERY HARD TO ACHIEVE

IT'Z ALL ZHE ZAME, OKAY!!!

THE FRENCH BAKERY

5.4. BREAD IMPROVERS

To make bread quality less dependent on the variations in raw ingredient quality and processing conditions during kneading, rising and baking, bakers add so-called bread improvers to their dough. The chemical bread improver potassium bromate was used until the beginning of the '90s, when it was banned because of potential carcinogenic properties. Now ascorbic acid, better known as vitamin C, is used with a complex mix of other substances such as emulsifiers, gluten-reducing agents, sugar, milk solids and a combination of enzymes. Since the early 1990s enzymes in particular have been used increasingly in the bakery.

5.5. ENZYMES

The addition of extra enzymes to the dough has the following benefits according to DSM[28]:

- *Improved dough handling and process tolerance*
- *Increased baked volume*
- *Finer crumb structure*
- *Improved crispiness and colour*
- *Softer crumb and extended shelf life*
- *Replacement of traditional emulsifiers*
- *Reduced reliance on high-cost ingredients such as gluten*
- *Acrylamide reduction (Textbox 5.3)*

True enough, enzymes have long been used in malting and baking, but that was in the form of malt flour and malt extract. These ingredients are also subject to strong variations in quality, so these days bakers prefer to use well-defined enzyme preparations. The use of α-amylases (enzymes that hydrolyse starches) derived from moulds began in the '60s. The α-amylases (nowadays produced from bacteria) produce dextrins (intermediary product in conversion of starch to sugars) from starch. These are further broken down into sugars by the naturally occurring β-amylases in dough. This improves the yeast fermentation, thus the rising process, and consequently the volume, crust colour and shelf life of the bread. A specific example with unexpected benefits is described in Textbox 5.4. Variously sourced proteases (enzymes that break down

[28] www.dsm.com/le/en_US/bake/html/role_enzymes.htm

It has been confirmed that a wide range of cooked foods – prepared industrially, in catering, or at home – contain acrylamide at levels between a few parts per billion (ppb, μg/kg) and in excess of 1000 ppb. This includes staple foods like bread, fried potatoes and coffee as well as specialty products like potato crisps, biscuits, crisp bread, and a range of other heat-processed products. Immediately following the initial alarming announcement at the start of the century, the food industry within the EU took action to understand how acrylamide is formed in food, and to identify potential routes to reduce consumer exposure. From the onset of the acrylamide issue, the efforts of many individual food manufacturers and their associations have been exchanged and coordinated under the umbrella of the European Food and Drink Federation (CIAA), to identify and accelerate the implementation of possible steps to reduce acrylamide levels in foods. These efforts are also intended to explore how the knowledge developed by industry might also be applied in home cooking and catering which contribute to more than half of the dietary intake of acrylamide. Applying the enzyme asparaginase in food in order to reduce acrylamide has been identified by various institutions as one of the solutions. PreventASe™ is the first asparaginase enzyme that is used in a commercialised product (DSM). Since October 2007 consumers in Germany have been able to buy a Christmas biscuit produced with PreventASe™.

The enzyme basically converts one of the precursors of acrylamide, asparagine, into another naturally occurring amino acid, aspartate. As a result, asparagine is not available anymore for the chemical reaction that forms acrylamide when carbohydrate-containing foods are heated. The PreventASe™ enzyme essentially reduces the formation of acrylamide, by up to 90%. PreventASe™ is not required to be listed on the product's food label, and requires no registration in most European countries (except for France and Denmark) as it is considered a processing aid. A safety record for review has been submitted to the French food safety authority AFFSA - resulting in an approval for the product. PreventASe™ has also been approved in Denmark and Switzerland. Also in the US, PreventASe™ can be applied without any further registrations. The FDA reviewed the DSM safety data and provided GRAS notification for PreventASe™.

Commission recommendation 2007/331/EC of 3 May 2007 on the monitoring of acrylamide levels in food required the Member States to monitor annually in 2007, 2008 and 2009 the acrylamide levels in certain foodstuffs, e.g. bread, potato crisps, instant coffee, etc. At the time of finishing this chapter, July 2010, the results of 2008 had just been published in a scientific report[30] by EFSA. The report in general suggests lower acrylamide values in 2008 compared to 2007, but soft bread, bread not specified, infant biscuit, and biscuit not specified, showed statistically significantly lower levels. Whether this represents a trend towards lower acrylamide levels over time should become clearer from the reports in the coming years.

[29] www.ciaa.be/documents/brochures/ac_toolbox_20090216.pdf
[30] www.nbc.nl/files/EFSA%20rapport%20acrylamide%20monitoring%202008.pdf

proteins) are normally used to reduce the elasticity of the dough in hard wheat varieties. Hemicellulases (or cellulases and pentosanases), whose purpose is to break down hemicellulose, are used not only to improve the baking properties of robust rye flours, but also to further optimise the dough properties and the quality of wheat bread. A multitude of other enzymes are also used in bakeries. A recent development, for example, is the addition of a new lipase (fat-hydrolysing enzyme) as the first enzymatic alternative for traditional emulsifiers, and there are still more in the pipeline. There's nothing wrong with traditional emulsifiers, except that they have an E-number. Food without E-numbers is seen to be more 'natural' and sells better. Since enzymes are 'natural' and have the same effect as an emulsifier, the latest trend is to add this sort of enzyme. Here too, however, modern biotechnology is beginning to play an increasingly important role, and the next question will be: is it still 'natural'?

5.6. RECOMBINANT ENZYMES

In late 1992 the Dutch Consumer's Association drew attention to the fact that much of our daily bread was made using bread improvers that contained an enzyme produced with genetically modified bacteria. This was confirmed by one of the biggest manufacturers of bread improvers, which means that this enzyme was already on the market before the Netherlands introduced legislation in July 1993 on so-called "novel foods". By 1997 five recombinant enzymes for use in bread-making were authorised by this legislation. According to the Association of Manufacturers and Formulators of Enzyme Products (AMFEP), a European industrial association set up in 1977, the following baking enzymes with genetically modified microorganisms were already being made and used in 1996: α-amylase, glucose oxidase, hemicellulase, lipase, malt amylase, protease, pullulanase and xylanase. AMFEP is a non-profit trade association which has taken a clear-cut and public stance on modern biotechnology since 1995. Textbox 5.5 gives their policy declaration at the turn of the century. On their very informative and up-to-date website you can find out all about enzyme regulations in the EC and their latest fact sheet on protein engineered enzymes[31]. They conclude this fact sheet with: "Protein engineering is regarded by AMFEP as a safe and useful tool in the development of improved enzyme products and processes that bring real benefits to manufacturers, consumers and society." AMFEP promotes an open dialogue on the use of this technology. Most of the bigger enzyme producers are full (14) or associate (9) members of AMFEP.

On the issue of complete safety in compliance with internationally accepted standards, various national and international expert committees have issued guidelines on how safety assessments should be conducted. These guidelines are all developed from the basic premise that the enzymes used in the processing of foodstuffs are per se intrinsically safe and that the analysis should focus on impurities and by-products, originating from the raw materials or produced during fermentation. These guidelines also apply to the safety assessment of enzymes produced with genetically modified microorganisms.

[31] www.amfep.org/papers.html

Bread enzymes are also good for the environment.

The Danish company Novozymes is one of the world's biggest enzyme manufacturers. The December 2005 issue of their newsletter, BioTimes, contained an interesting article about the added benefits of Novamyl, one of their registered amylases, often used in bread making. These added benefits have been established by means of a so-called 'Life Cycle Assessment' (LCA). LCA is a methodology that enables a comparison to be made of the effects on the environment of alternative production technologies that have the same user benefit. LCA takes a holistic look at the business and inspects the whole production system, from the manufacture of raw materials to the disposal of waste. ISO guidelines ensure that LCAs are performed in a transparent and standard way.

The addition of Novamyl enhances the taste and texture of breads and also produces a delicious, fresh bread with a long shelf life (10 to 14 days). This long shelf life enables bakers to use their production facilities more efficiently. There is less need for them to go from one product to the next and they can therefore prolong production runs. An LCA demonstrated that besides considerable reductions in energy consumption and greenhouse gas emissions, there was also a

reduction in the cost of transport of bread from bakery to shop. Approximately 45 percent of the reduction in energy consumption is a consequence of this reduction in transport. The savings have also led to savings in agricultural production. Less bread is wasted, so less grain is needed, so less fertiliser is used, meaning less acidification of the soil. The LCA applies primarily to the American situation, but there is a clear message for the EU, where many member states have a strong preference for crusty breads like baguettes with a much shorter shelf life.

AMFEP's policy statement is designed to prevent any misunderstanding about what people want and what people do. AMFEP members therefore ensure that the enzymes used in the processing of foodstuffs are obtained with non-pathogenic and non-toxic microorganisms, i.e. microorganisms that have a 'clean' safety record, with no reported cases of pathogenesis or toxicity that can be ascribed to the microorganism in question. The raw ingredients that are used for cultivating the microorganisms are carefully selected so that they do not contain any components that are harmful to health. Every time a new microbial strain is developed with improved enzyme production capacity or the production conditions are changed, the potential impact on safe usage is carefully evaluated on a case-by-case basis. Every new strain is checked for its primary taxonomic properties. If the production strain contains recombinant DNA, the properties and the safety record of the donor organisms that delivered genetic information for the production strain are analysed. The safety of the product is usually backed by documentation from toxicological safety studies. Consistency and quality are assured by production

under GMP (Good Manufacturing Practice). All this and more ensures that enzymes are safe and can be safely used. It is therefore hardly surprising that the JECFA (Joint Expert Committee on Food Additives of the FAO/WHO) concluded that there is no need to limit exposure to enzyme preparations that they have evaluated, also with respect to allergy (Bindslev-Jensen, Skov, Roggen, Hvass, & Brinch, 2006); none of them have been allocated an ADI (Acceptable Daily Intake). You can read more about the authorisation, labelling and traceability of these enzymes in the section on legislation.

5.7. TRANSGENIC CROPS

The advent of genetically modified (transgenic) soya has revived interest in the use of modern biotechnology in bakeries. Soya ingredients are used in abundance in the bakery industry: processed soya beans, soya flour, soya oil and lecithin. Only the protein-containing fractions of the soya ingredients can carry the characteristics of genetic modification, since the DNA determines which proteins a cell can make. Soya oil and lecithin are products that don't contain protein, and will therefore not usually be altered by the modification. Bread rises because large protein molecules in wheat dough form a network, i.e. gluten, that gives the dough its strength, elasticity and capacity to expand, enabling the dough to trap the carbon dioxide gas produced by the yeast. It is primarily the large gluten proteins that bind together during the mixing and kneading process to form much bigger polymers. It is for this reason

GOOD MANUFACTURING PRACTICE

THIS GRAIN PARTICLE IS OKAY! ONLY 13,786 TO GO ...

Part 2: Our daily food and drink

TEXTBOX 5.5.
AMFEP's policy declaration on modern biotechnology.

1. AMFEP fully supports and **is committed** to continuing the use of genetic modification (modern biotechnology) for the development of improved enzyme-producing microorganisms. This technique offers a whole range of benefits and it is important that it is researched with a view to its use by society as a whole. Genetic modification should be regarded as a logical extension of traditional genetic techniques.

2. The microbial enzymes that are produced by AMFEP members are used in a wide range of industrial applications. The introduction of genetic modification can offer the following **benefits** with regard to the production and/or quality of these enzymes:
   - greater production efficiency and thus less use of energy and raw materials, and less waste;
   - availability of enzyme products that for economic, enviro-technical or, as regards production staff, health reasons would otherwise not be available, making new applications possible;
   - technical improvements due to higher specificity and purity of enzyme products.

   - for these reasons we see genetic modification as an extremely important tool in the production of our enzymes.

3. AMFEP is of the opinion that all enzyme products must be judged on their intrinsic properties and not on the basis of the method used to develop the production organism. Our products – whether or not they are produced using genetically modified organisms – are only put on the market once complete **safety** has been established according to internationally accepted norms.

4. AMFEP believes that the '**right to knowledge**' and the '**right to choice**' of the **consumers** must be respected and that an open dialogue is the way to win their trust in modern biotechnology. Therefore, the AMFEP members are prepared to support their clients and actively inform them if an enzyme is or is not produced using genetically modified organisms.

5. AMFEB members will continue to provide clients with enzymes made using genetically modified organisms and to help them tackle consumer concerns. Members will **not compete** on the basis of this technology by using claims that support or refute genetic modification.

---

that so much time and effort is spent on selecting and breeding wheat varieties with a high gluten content. This is a complex procedure as far as genetics is concerned, because six genes are involved. Despite that, in 1997, Australian and British researchers showed that genetic modification has huge potential in this area too (Barro *et al.*, 1997). Compared to dough made from non-transgenic seeds, the dough from seeds with one or two extra gluten genes showed a proportional increase in strength and elasticity.

Researchers in the US are also working on transgenic wheat with higher gluten levels. In 2004 and 2005 trial fields were set up in California and Idaho with a transgenic variety developed by the USDA (United

States Department of Agriculture). In 2006 baking tests were conducted with flour from the first harvest in Kansas. Although the results were somewhat mixed, USDA researcher Ann Blechel concluded nevertheless that they had succeeded in producing a wheat variety that would finally give bakers a competitive edge (Anonymous, 2006). However, the question remains as to when bread made from this sort of flour will be sold over the counter to consumers.

The American Bakers Association (ABA), which represents 85% of the major bakeries in the US, would also like to see modern biotechnology being used to develop flour with increased levels of vitamins, reduced caloric values, fewer allergens, etc. To help broaden acceptance, a new GM wheat needs to include nutritional improvements for consumers and/or improved milling and baking characteristics, according to Hayden Wands, director of procurement at Sara Lee Corp and an official of the ABA. "We are not one hundred percent convinced that our customers will go for a GM wheat unless it has enhanced characteristics," Wands told a gathering of representatives from agriculture and the technology industry at the Biotechnology Industry Organization convention in Chicago (Gillam, 2010). According to a researcher from the University of Melbourne (Bhalla, 2006), wheat with improved characteristics (Table 5.1) is expected to have a substantial effect on food security and on our society in general. Bhalla believes that progressive and consistent implementation of transgenic crops is a basis for an increase in productivity, which has environmental and economic advantages for both growers and consumers.

In the August 2007 issue of *Nature Biotechnology* researchers from Pioneer Hi-Bred, a subsidiary of DuPont, published an article declaring that they had identified a key gene in phytic acid biosynthesis. By using special genetic techniques to deactivate this gene (gene silencing) in seed tissue from corn, they were able to create a maize variety with seeds containing very little phytic acid, but high concentrations of inorganic phosphate. This is good for feed given to cattle with one stomach, such as chickens and pigs (see also Chapter 7 on meat). The same genetic technique also seems to work in soya beans, suggesting that it might also work in other crops. If that is the case, the ground has been cleared for producing a grain variety with an improved dietary value.

By 2004, the Monsanto Company, a leader in the production of seeds for genetically engineered crops, had made substantial progress in the development of GM wheat varieties for North America. However, suddenly in that year, the company scrapped its wheat program, in part because of opposition from North American grain merchants and growers, as well as concerns that some major foreign importers would reject imports of all American wheat because they could be contaminated with genetically engineered varieties. In their opinion paper, Miller and Carter (2009) plead for a return to this technology. According to these authors, greater productivity in wheat farming achieved with improved varieties would confer an important environmental dividend: wheat is the largest crop in the world in terms of area cultivated (220 million hectares) and is the second largest irrigated crop (each bushel produced requires about 40,000 litres of water on average; it is three times thirstier per bushel than maize for example); therefore, enhanced productivity would conserve both farmland and water. They conclude their paper with: "Monsanto and the United States wheat industry might already have been relegated to the position of second mover, and whoever wins the race to produce desirable genetically engineered wheat varieties to the marketplace will enjoy a strong cost advantage and attract market share in many importing countries. Agriculture remains an important American industry; one that should have learned by now that, if it is slow to bring the best technology to the table, other countries will eat our lunch." The authors can be happy again. In *The Wall Street Journal* of 7 July 2010 Ian Berry reports that the world's largest seed maker Monsanto and the German chemical giant BASF are starting to develop genetically modified wheat again as part of an expanded joint venture (Berry, 2010). The declining production in the US has sparked renewed farmer interest in developing a stronger variety of wheat.

5.8. LEGISLATION

Chapter 2 describes the role of the European Food Safety Authority (EFSA) as the cornerstone of EU risk analysis concerning food and animal feed safety. The

Table 5.1. Examples of transgenic properties that may help address the needs of a growing world population.

Stress tolerance
  Abiotic stress
  - Drought tolerance
  - Salt tolerance
  - Oxidative-stress tolerance
  - Improved tolerance for aluminium, boron, cold and heat
  Biotic stress
  - Resistance to pathogenic fungi, viruses and bacteria
  - Resistant to insects and nematodes

Agronomic properties
  - Herbicide tolerance
  - Improved efficiency in water use
  - Hybrids

Quality properties
  - Improved grain quality
  - Improved nutritional quality – lower phytate levels (Raboy, 2007), higher macronutrient content, greater essential micronutrient content, and better amino acid composition
  - Modified gluten composition for people suffering from gluten allergies
  - Special types of wheat with health-promoting nutraceuticals in the grains

authorisation, labelling and traceability of genetically modified organisms, foodstuffs and animal feed are dealt with by two EU directives, 1829/2003 and 1830/2003, which have been in force since 2004. Not all ingredients involving genetic modification need to be labelled as such; recombinant enzymes, for example, which are used as processing aids, do not as yet need to appear on food labels. However, this legislation is still in the development stage.

EU regulation no. 1829/2003 imposes tighter rules on the authorisation (safety assessment and permit allocation) and the labelling of GMOs and genetically modified food and feed. Genetically modified additives (substances intended to make a product look better, last longer, be lighter, etc.) and flavourings also fall under this regulation. Genetically modified food is food that consists wholly or partially of genetically modified organisms or is produced with them or contains ingredients that are produced using GMOs. This applies regardless of whether it can be demonstrated that DNA or protein created by genetic modification exists in the end product. The reasons for this are that the consumer must be in a position to make a well-informed choice between traditional and genetically modified food.

Since 18 April 2004 this regulation makes it obligatory to label genetically modified food and feed as such, when it is delivered to the end user (consumer) or to an institution. As far as unpackaged products or very small packages are concerned, the information about the presence of genetically modified components must be placed permanently and visibly on or immediately next to the sales shelf. This must be done in a font size that is big enough to be easily identifiable and legible.

CHOOSING BREAD IS GETTING MORE AND MORE DIFFICULT

In certain cases special features or properties must also be mentioned on the label, especially if, due to the genetically modified component, the food has a different composition, a different nutritional value or nutritional effect, a different use or certain consequences for the health of certain population groups, compared to the same food without the genetically modified component. A notification is also obligatory if a food may lead to ethical or religious objections.

A tolerance value (0.9%) has been set for ingredients that contain traces of genetically modified organisms due to unforeseen contamination (during cultivation, harvesting, transport or storage). In order to be able to demonstrate that the presence of the GMOs is unforeseen, companies must be able to produce proof to convince the authorities that they have avoided the use of GMOs and that it is therefore a case of unintentional contamination. GMOs that are (still) not found to be safe may obviously not be present in the product (zero tolerance).

EU regulation 1830/2003 guarantees the availability of the relevant information concerning genetic modification in all phases of the marketing authorisation of GMOs and the food and feed produced with them. Information on the presence of GMOs must be conveyed in writing at each stage of the chain from "farm to fork" and kept for a period of five years. Suppliers and purchasers (excluding consumers of course) must therefore also be known. If the supplier doesn't provide any information, the regulation stipulates that there is no obligation on the part of the purchaser to mention GMOs on or next to the product.

As yet, technical agents such as enzymes do not need to be labelled as GMOs. Nor do milk, meat or eggs from animals that have been fed GM feed need a GMO label. The same applies to substances that are produced by fermentation using genetically modified organisms (e.g. certain additives or vitamins), but where no residues of the microorganism appears in the ingredient (contained use). N.B. There is a new regulation which is waiting to be approved by the EC Council, which will introduce changes to these exemptions.

If, during an inspection, it appears that a genetically modified ingredient has been used in a food without it being mentioned on the label, this will be sufficient reason to take legal action.

5.9. IN CONCLUSION

In the (near) future there are likely to be an increasing number of bakery ingredients originating from transgenic plants; examples of which are wheat, potato and sugar beet, alongside the currently available soya and corn. We are also seeing a rise in the use of enzymes produced with recombinant microorganisms as bread improvers. In consultation with organisations such as Commodity Boards and AMFEP, EU governments are working hard on authorisation and labelling rules that are fair and acceptable to all parties. In the end these must enable the consumer to choose between genetically modified or non-genetically modified food. This choice will also result in more pressure on growers and sellers to fully separate (make traceable) product flows. The growing application of modern biotechnology in food production and the relatively rapid changes in legislation in this area, also mean that the bakery industry must keep pace with and, in particular, capitalise wisely on changing circumstances, especially as concerns public opinion. It's good to see national organisations helping out in this regard by translating the directives into real workplace language and finding user-friendly ways of communicating this information to bakers.

We would like to conclude this chapter with a magical quote made at the turn of the millennium by the American culinary writer John Thorne: "Bread is an unparalleled and key source of nutrition among foods. The baked dough feeds the body, but the dough itself must be fed by the baker, and the process of preparing and baking offers a kind of intellectual and psychological nourishment." Keep that in mind!

5.10. SOURCES

Anonymous. (2006, 10 July). Gluten in rich wheat trials. *AgraFood Biotech*.

Barro, F., Rooke, L., Békés, F., Gras, P., Tatham, A., Fido, R., *et al.* (1997). Transformation of wheat with high molecular weight subunit genes results in improved functional properties. *Nature Biotechnology, 15*(12), 1295-1299.

Berry, I. (2010, July 7). Monsanto, BASF Turn Attention to Wheat. *The Wall Street Journal*.

Bhalla, P. (2006). Genetic engineering of wheat-current challenges and opportunities. *Trends in Biotechnology, 24*(7), 305-311.

Bindslev-Jensen, C., Skov, P., Roggen, E., Hvass, P., & Brinch, D. (2006). Investigation on possible allergenicity of 19 different commercial enzymes used in the food industry. *Food and Chemical Toxicology, 44*(11), 1909-1915.

Gillam, C. (2010, May 4). US millers, bakers urge caution in GMO wheat work. *Reuters*.

Miller, H., & Carter, C. (2010). Genetically engineered wheat, redux. *Trends in Biotechnology, 28*, 1-2.

Raboy, V. (2007). The ABCs of low-phytate crops. *Nature Biotechnology, 25*(8), 874-875.

WINE: ONE OF THE OLDEST BIOTECHNOLOGICAL PRODUCTS

"In vino veritas"

Wine is probably the oldest of all biotechnological products, and yet modern biotechnology offers a whole range of possibilities for its production. Every year approximately 27 billion litres of wine are made from grapes plucked from about 8 million hectares of vineyard. The "magic" world of wine is currently experiencing a real revolution with its transformation from a production-oriented to a market-oriented industry. And this revolution depends on innovations in the area of modern biotechnology! Some of these will be discussed in this chapter.

WINE, ONE OF THE OLDEST BIOTECHNOLOGICAL PRODUCTS

6.1. WHAT IS WINE?

According to the shortest possible definition, wine is fermented grape juice. According to Wikipedia, wine is a beverage produced when the juice of grapes is fermented. Needless to say, the International Organization of Vine and Wine (OIV) has a little more to say on the subject[32]. According to the OIV's "International Code of Oenological Practices", wine is the beverage resulting exclusively from the partial or complete alcoholic fermentation of fresh grapes, whether crushed or not, or of grape must (juice). The Wine & Spirit Trade Association has also developed a similar standard definition[33]: Wine is an alcoholic beverage, obtained by fermenting freshly picked grapes, the fermentation of which takes place according to the local traditions and practices in the area of origin.

However, John Baldwinson says in *Plonk and Superplonk* (1975) that there is something missing from the above definitions. Nothing is said about the pleasures of wine: the complex colours, tastes, aromas, associations. Nothing about the glow a good wine can give or about the natural (some say lively) character. Nothing about it being good for you, but above all, there was nothing about the fact that wine can make you happy. "Wine as a panacea for unhappiness." The pronouncement by Euripides' Bacchae is completely in agreement on this point: "Where there is no wine, there is no love, or any other pleasure left for men."

[32] www.oiv.int
[33] www.wsta.co.uk

Scientific proof of the benefits of wine (for the heart and blood vessels) when consumed in moderation, was only demonstrated at the end of the 20th century. More than a century earlier the first oenologist (Section 6.4), Louis Pasteur (1822-1895), came to the same conclusion:

"Wine can be considered with good reason as the most healthful and the most hygienic of all beverages."

6.2. THE FIRST WINE

Since time immemorial humans have been getting microorganisms to work for them. There are indications that wine was already being made from grapes about 8000 years ago. Chinese rice wine was made during the Shang dynasty between 1600 and 1100 BC. There's no doubt that there were several "inventors" of wine. However, history names no names, unless Noah can be awarded this accolade.

Vitis vinifera is the grapevine cultivar most commonly used for producing wine grapes and is native to the Caucasus. Its geographic central position meant that this vine quickly spread to the rest of the world. Until 1991 it was believed that wine making was first performed approximately 3000 BC. In 1991, however, Virginia Badler of the University of Toronto presented her research results at a conference of wine experts in California. She had examined a Persian earthenware amphora from 3500 BC which had a red stain on the bottom. Infrared spectroscopy revealed that the stain contained tannin and tartaric acid among other things. Both substances occur in wine. Previously, scholars from Israel had discovered that grapes had indeed been grown there in 3500 BC. The difference between wild and cultivated grapes is easy to identify by the different forms of grape seeds. So clearly wine was already made 500 years earlier than thought until 1991. Vinicultural knowledge spread further afield as a result of trading with neighbouring countries, arriving eventually in Egypt. At the same conference it was revealed that in 3000 BC wine jugs indeed existed in Egypt. The ancient Egyptians have left behind many images and other evidence of wine dating back to that time. The Egyptians themselves did not grow grapes, so the wine must have been imported. At that time not only was the art of winemaking already established, but there were thus also a lively trade in wine. In Mesopotamia, in the present area of land in Iraq that lies between the Euphrates and the Tigris, 7000-year old pitchers containing traces of wine have meanwhile been found in archaeological digs.

Grape vines reached Greece about 2000 BC. Amphoras (Textbox 6.1) and a winepress found on Crete date back to 1500 BC. The art of winemaking then spread from Greece to Italy, France and Spain. These three countries were once the biggest producers in the world, until the United States, Argentina, Chile, South Africa and Australia joined them.

TEXTBOX 6.1.

The amphora and traditional Greek wine: retsina.

An amphora is a jug, with two handles, that narrows to a point at the bottom. Amphoras were generally used at that time to store and transport liquids like oil or wine, but also to store grain. Apart from its use as a name for a jug, a Greek amphora was also a measure of volume for liquids (approximately 23 litres). Since an amphora was too porous a vessel in which to keep wine, resin was added to the wine to "seal" the amphoras. Because resin also vastly prolonged the shelf life of wine, the Greek viniculturists continued to add resin to their wines. The Greeks got used to the resin taste and even became attached to it, whereby retsina became the most well-known and best-loved Greek wine and is still made in abundance. As an outsider you either hate it or you love it.

The entire wine industry is presently noticing the (negative) effects of global warming. It seems probable that this global warming will shift the wine boundary further and further north. Winegrowers in the three biggest European wine countries, France, Italy and Spain, are already moving to cooler areas and in the Netherlands a growth in the number of commercial vineyards is clearly visible. Maybe modern biotechnology has the solution to the problems created by this climate change[34] (Section 6.9).

6.3. ALCOHOL AS A STIMULANT

In the previous chapter on bread we described how yeast produces carbon dioxide when the yeast cells grow on the sugars in the dough. There is, however, another by-product, albeit in relatively small quantities, because the bread yeast strains used are mainly selected for their capacity to produce carbon dioxide. For most other applications, that by-product - alcohol - is much more interesting and is even the main component. This 'stimulant' was probably discovered back in prehistoric times, as mentioned above, when people began making a sort of wine and beer. It was initially produced domestically, but later developed on a semi-industrial scale in sheds behind inns and public houses, and in monasteries. Much later still it was produced on an industrial scale in wineries and breweries. What's certain is that the first winemakers and beer brewers had no idea how or why these fermentation processes took place. The

best they could come up with was a semi-magical explanation. With the scientific knowledge we possess today about how complex these processes are, we can only sit back and admire these so-called "primitive" people. Time and time again they managed to develop something with which they were not only able to make wine in a reasonably reliable way, but also other biotechnological products like cheese, bread, beer and soft leather. All without any scientific understanding of the processes.

6.4. THE SCIENTIFIC DISCOVERER: LOUIS PASTEUR

Scientific understanding of fermentation processes really only started in the second half of the nineteenth century when in 1867 Louis Pasteur discovered a number of undesirable microorganisms which spoiled the fermentation of wine and beer. In fact, microorganisms had been discovered a few hundred years earlier in 1676 by Antonie van Leeuwenhoek. This Dutch scientist had designed a primitive microscope and studied what he himself called 'the smallest animals I have seen thus far'. Pasteur was asked by Napoleon III to find out why wine, already an important export item for France at that time, spoiled during transportation to consumers abroad. His studies resulted in three simple, but oh so very important, guidelines for making wine, and for fermentation in general.

The first guideline, which now seems so obvious to us, is hygiene. Wine can only be prevented from

[34] www.gmo-compass.org/eng/database/plants/73.grape_vine.html

rapidly turning into something else, usually vinegar, if it is produced and stored in clean containers, as well as being sealed off from air and thus from the many microorganisms floating around in it. Interestingly though, making vinegar from wine is also an old biotechnological process using acetic acid bacteria.

The second guideline is the use of one of Pasteur's innovations, namely the process that bears his name, *pasteurisation*. This process is based on the necessary hygiene procedure for extending shelf life. When wine, beer or other products like milk, are heated and stored for a while at 75° C, most of the pathogenic and food-spoiling microorganisms are killed, so reducing the possibility of a loss of quality, and spoilage, as a result of undesirable fermentation.

WINE INNOVATION: PASTEURISATION

HEY DAD, LOOK! I'M EXTENDING THE SHELF LIFE OF YOUR WINE!

Pasteur's third scientific contribution consisted of understanding the importance of oxygen in the fermentation process. By growing yeast without sufficient oxygen for the complete conversion of sugars to carbon dioxide and water, alcohol is primarily created. So, an abundance of oxygen stimulates the growth of the cells and reduces alcohol formation.

6.5. HOW IS WINE MADE?

Wine is the product of fermented grape *must* or juice. The diagram below shows a very simple overview of the red and white wine production routes. However, it is important to realise that many hybrid forms and specific details have been introduced by knowledge, art and tradition. Harvested grapes are processed immediately. Crushed whole grapes, called must, are used at the beginning of the fermentation process for making red wine. Colourings, aromatic substances and flavourings in the skins are extracted in this process. White wine is produced by first removing the skins in a press before the fermentation. Because the colouring substances are present in the skin, white wine can also be obtained from red grapes. Thus only juice is further processed, starting sometimes with clarification. The fermentation can be set in motion by inoculation with a commercial strain of the yeast *Saccharomyces cerevisiae*. It is also possible to conduct a natural or non-inoculated fermentation by using the wine yeasts that are present naturally on grapes or that are airborne via the natural flora. At the start of such a natural fermentation, there are also many wild, non-*Saccharomyces* yeasts and bacteria present, but by the end of the process *Saccharomyces cerevisiae* has outnumbered them.

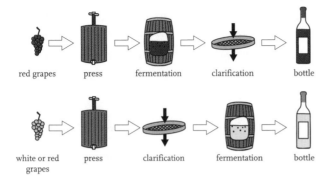

red grapes   press   fermentation   clarification   bottle

white or red grapes   press   clarification   fermentation   bottle

Aside from climate, grape and yeast variety, and soil type, the metabolic activity (metabolism) of other microorganisms contributes significantly to the taste and aroma of the final wine. Lactic acid bacteria, for example, play an important role in the formation of flavours and aromas. These bacteria convert the sharp-tasting malic acid (malate) in the must into much milder lactic acid (lactate) and carbon dioxide. This process is called malolactic fermentation and is responsible for the "fatty" character of the better wines. This conversion is also especially important in wines that would otherwise be too acidic.

After fermentation, wine undergoes a number of further minor or major processes. Red wine first goes through the press to remove the skins and other solids, and is clarified if necessary. Many wines are then left to mature before being bottled. That can be for a short time in large tanks, or for years in oak barrels, and many other configurations in between. During this time all sorts of chemical changes occur that increase the complexity of the wine, i.e. the bouquet, the flavour becomes more diverse. The wine can be processed further with adsorbing substances, such as proteins, in order to remove undesirable substances that cause cloudiness, bitterness or acidity.

Depending on the winery and the quantity of fermentable substances still present, the wine sometimes undergoes microfiltration so that it enters the bottle sterile. As Pasteur established, this is important in terms of shelf life. However, it should be noted that generally the wine does not undergo a heat treatment. This would destroy too many aromas and flavours and would seriously damage the character of the wine. Good hygiene is the watchword.

6.6. ENZYMES ARE THE SOLUTION!

Over the last few decades enzyme manufacturers have introduced a series of enzymes onto the market for all manner of applications in wine production. There are, for example, enzymes called pectinases and hemicellulases that stimulate the extraction of juice during the crushing and pressing. The cell walls of grapes consist largely of pectin and hemicellulose which are broken down by these enzymes. They dissolve, as it were, so that the fruit juice in the cells is more easily released.

Enzymes are also added for a more efficient release of flavours and aromas, thus allowing the characteristic aromas of grapes to develop to their maximum potential. This is done primarily with the aim of increasing the concentrations of free terpenes in wine. So, although you can't make a bad wine good, you can make a good wine better. These enzymes provide the winegrower - the most traditional of all craftsmen - with all kinds of new tools for enhancing the quality, consistency and stability of his product.

A classic problem in winemaking is the clarification, i.e. making the wine clear. After every harvest winemakers are confronted with the issue of whether the wine will become clear and therefore be easy to filter. The cloudiness and accompanying filtering problems are chiefly caused by the presence of pectins and glucans. These are large molecules that are present in the wine forming solid aggregates which result in a cloudy wine. The gummy glucans in particular can make filtering problematic. Some of these glucans are associated with the *Botrytis cinerea* mould, which makes the grapes decay. Even if only a very small percentage of the grapes used is partially rotten, this can make the resulting wine incredibly difficult to filter. The wine can remain cloudy for weeks, even months, which is detrimental to the quality. Enzymes can offer a solution here too. By adding a mixture of pectinases and glucanases immediately after alcoholic fermentation, the pectins and glucans are broken down and the wine quickly clarifies, thus eliminating filtration problems.

Many of the enzymes used are products of modern biotechnology, because they are made with recombinant microorganisms. The enzymes themselves are authentic, i.e. no different from those produced by the original organisms. Furthermore, since they are used in relatively small quantities, as auxiliaries, there is no obligation to mention them on labelling (this may change in the future; see previous chapter on bread). Some of these enzymes, and others too, are also used on a large scale to make fruit juices.

6.7. CHAMPAGNE WITH A FLICK OF THE WRIST

Champagne is the most prestigious of all the sparkling wines. It is made according to the quality control rules of its *appellation*: in a very limited region, from specific grape varieties and using very well-defined procedures. The production process for sparkling wines usually consists of two main steps. First, a basic wine is made. This has certain specific properties, for example, a moderate alcohol content by using early plucked grapes which still contain very little sugar. After the mixing of various basic wines, if this is required, a second fermentation is initiated. For champagne this is done in the bottle that the consumer purchases. This second fermentation is brought about by the addition of the *liqueur de tirage*, a sugar and yeast mixture. The bottle is then fitted with a crown cap and stored horizontally at 11-12 °C. The second fermentation process only takes a few months, but the champagne is then left to mature for two to eight years, or even longer for top champagnes.

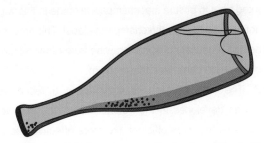

The next stage is the *remuage*: the turning of the bottles in special racks to collect the yeast deposits in the bottle neck at the crown cap. The bottles are slightly turned with a rapid motion several times a month and returned to an increasingly vertical position, with the neck pointing downwards. The length and intricacy of this procedure is determined by various factors, in particular the settling properties of the yeast. A month at least is usually required. The duration, the storage, and the labour-intensive nature of the remuage process determine the cost. So a great deal of effort has been put into simplifying this step, including automation.

Ten years before the end of the last century, an ingenious method was devised by the famous champagne house Moët et Chandon. Live yeast was 'immobilised' in little 1 mm balls. The immobilisation process that they designed for this purpose is such that the immobilised yeast behaves in the same way as the normal, free yeast. However, now after fermentation in the bottle, the *remueur* simply has to turn the bottle upside down and the yeast balls roll along to the crown cap. In short, with the flick of a wrist the process is complete, resulting in enormous cost savings. And yet, as far as we know, this process is still not used for making champagne. The employee unions, who obviously fight for employment opportunities, have worked hard to ensure that champagne made in this way is not allowed to carry the champagne label. The process is sold to producers of sparkling wine from other regions (Spain) which don't have this problem.

N.B. The immobilisation of enzymes is a technique that was developed in the 1960s and 1970s, in order to facilitate the more efficient and more effective use of these biocatalysts. Immobilisation of entire cells followed in the 1980s. Now many of these immobilised biocatalysts are used on a large scale in industry.

Nowadays the rest of the process, i.e. the disgorging, the removal of yeast, refilling, and recorking, is automated almost everywhere. For the disgorging, the bottles, with the crown cap still pointing downwards, are inserted a little way into a low-temperature bath so that only the few millimetres of champagne containing the yeast in the neck are frozen. The bottle is placed upright and the crown cap with the frozen champagne/yeast prop is removed. The space left is topped up with the *liqueur d'expédition*, which is not just a straightforward sweetener, but a method of further improving the champagne. The quality of this liquid, the subsequent maturation, the character of the wine, the quality of the sugar, and the recipe, are all important for the quality of the end product. Finally, the bottle is hermetically sealed with the special cork, the wire collar and the cap. The final result: a party!

6.8. MANIPULATION OF WINE YEAST

Winemaking may well be older than documented history, but there is still room for improvement in many areas of the process. This is where the application of modern biotechnology has so much potential, and why modern winemakers are so keen to find out exactly what it has to offer. There are research groups all over the world working on the genetic modification of yeasts in order to:

- *improve the settling of wine yeasts;*
- *optimise the balance between acid and alcohol levels;*
- *intensify the colour;*
- *make the wine "fuller" or "fattier";*
- *prevent undesirable substances entering the wine;*
- *increase the concentrations of health-improving components.*

Yeasts that easily flocculate and/or easily precipitate are worth their weight in gold for sparkling wines, as outlined above for champagne. Every yeast type has a set of genes that causes flocculation. But this set is often not switched on. At the INRA[35] (French National Institute of Agronomic Research) in Montpellier, researchers have made a recombinant champagne yeast with its own specific *switch* to activate the flocculation genes in these traditional yeast cells. In principle, therefore, it will be possible in the future for a winemaker to flocculate the yeast to order.

At the same institute, researchers also inserted genes from lactic acid bacteria into wine yeasts, enabling these to convert malic acid into lactic acid and carbon dioxide, i.e. perform the so-called malolactic fermentation without using lactic acid bacteria. Malolactic fermentation is often problematic, because lactic acid bacteria do not thrive well in alcohol. Alcoholic and malolactic fermentation can now take place simultaneously and can be executed by one and the same recombinant yeast. The use of this type of yeast strain has now been approved by the American FDA (Food and Drug Administration), and has been granted GRAS status (Generally Recognized As Safe). According to the literature, it is already being used commercially in Moldavia and the US. The yeast developed by Springer Oenologie (Textbox 6.2), part of the Lesaffre Yeast Corporation, can induce the alcoholic as well as the malolactic fermentation in a matter of five days; so saving the wine producers time (*AgraFood Biotech,* 25 June 2007, p. 7). According to recent research conducted by Canadian scientists,

wine produced with this genetically modified yeast has more of the desirable volatile acids and better colour properties than wines produced with traditional yeasts plus lactic acid bacteria. An analysis of volatile components and a sensory evaluation has shown that industrial production of wine with the recombinant yeast strain is suitable for the commercial production of quality wines.

BIOTECHNOLOGY CAN TURN POOR WINE INTO GOOD WINE

BUT CHANGING THE LABELS DOES THE TRICK AS WELL!

Although the name of the quoted writer in Textbox 6.2 is unknown to us, we largely agree with her/him. We also care a great deal about the future of wine, but in our opinion it will never be just another manufactured beverage. No two wines are the same, not now, not ever! Our line in the sand bans poor quality and unsafeness. That will guarantee a blooming enterprise, with no risk of destroying the whole venture.

There have been other fascinating attempts by biotechnologists to make good wine from poor wine using recombinant yeasts. Good wine tastes full-

[35] www.inra.fr

GM yeasts: the next battleground?

On the wineanorak website[36] it is predicted that the next battleground in the wine world will be the controversial use of genetically modified (GM) yeasts in winemaking. We quote from it:

"Plenty of these genetically modified strains already exist in laboratories around the globe, but they haven't previously been commercialized because of the negative reactions of consumers to GM food products. The scientists are busy engineering beneficial traits into wine yeasts even though they know they won't be useful for commercial winemaking for the foreseeable future … Now, however, a GM yeast strain, called ML01, has been commercialized and is authorized for use in the USA. This yeast, made by Springer Oenologie, has been the recipient of two extra genes (known as transgenes). The first is a malate transporter gene from another yeast, Schizosaccharomyces pombe, and the second is the malolactic enzyme gene from Oenococcus oeni, the main bacteria responsible for the natural malolactic fermentation that occurs in many wines after alcoholic fermentation. This yeast is therefore able to carry out malolactic fermentation (normally done by bacteria) at the same time as alcoholic fermentation. There are several advantages to this. The first is that processing wine becomes much faster. The second is that there is less risk of wine spoilage because there is no delay between alcoholic fermentation and the onset of malolactic fermentation,

a stage at which wine can be at risk. Also, the resulting wine is less likely to contain biogenic amines which are produced by the bacterial malolactic fermentation and which can have negative health effects. In the USA yeasts are classified as processing agents, and thus wines made with this yeast would need no declaration that they contained GM ingredients. This allows GM yeast to enter winemaking 'under the radar', with consumers or advocacy groups none the wiser. In many other countries, such as New Zealand and Australia, the regulations are more stringent, and yeast is considered as part of the ingredients of wine.

So is anyone making wine using this GM yeast? If they are, they aren't telling anyone, for understandable reasons. In response to the commercial approval of ML01 in the USA, the Australian Wine Research Institute has issued a statement declaring that no GM yeasts will be used in Australian wine for the foreseeable future. But because it is so much easier to produce yeasts with desirable properties by GM technology (and there are some traits that are impossible to select for by conventional breeding), research continues apace globally on GM yeast technology. So what's the big deal? Aren't GM microbes used all the time? … Supporters of the technology argue that what they are doing by developing GM yeast strains is not with the intention of creating fake wines, but with a view to unlocking the latent flavour and aroma potential of grape must by using yeasts with special properties. One yeast researcher has even gone on record as stating that the best wines are still to be made, and that this technology is one way forward. What do I think? As a scientist who cares a great deal about

[36] www.wineanorak.com/GM_yeasts.htm

the future of wine, I favour a cautious approach: if GM yeasts become widespread, the danger is that wine will be seen as just another manufactured beverage. If we kill the 'naturalness' of wine, we run the risk of destroying the whole venture. So although it rankles with me a bit to *knock the elegant science involved in engineering new wine yeasts, I'm afraid I'm going to voice my disapproval at the use of GM yeasts in wine. I think it's time we drew a line in the sand and banned the use of GM organisms in winemaking."*

---

bodied, a property that stems to a large extent from the glycerol in wine. However, the strict AOC rules do not allow glycerol to be added. But who can protest if a "new" yeast produces a little more glycerol in the wine? This yeast now exists, although a lot more manipulation is required, because it also produces quite a few undesirable by-products, such as acetic acid.

Saccharomyces cerevisiae, the wine yeast par excellence, can convert arginine, one of the most prevalent amino acids in must, into ornithine and uric acid during fermentation. However, *S. cerevisiae* does not immediately use all the uric acid produced. "Free" uric acid reacts spontaneously with the ethanol in wine to create a new substance called ethyl carbamate. Unfortunately, ethyl carbamate is carcinogenic. Research has shown, however, that an industrial wine yeast can be genetically modified so that the production of ethyl carbamate in Chardonnay is reduced by almost 90 per cent. Complete removal also seems feasible. In fact, Japanese scientists have already achieved this in *sake*.

Red wine contains polyphenols, including quercetin and resveratrol that come from the grape skins. These substances are attributed with favourable health effects, including protection against heart attacks. However, their real effectiveness in this area is still questionable, according to Katan[37], an expert in the field of human nutrition and the person who has for years explained to the general public what is fact and what is fiction in the claims made by the food industry. In his view, red wine is probably no better than vodka at protecting us against the risk of heart attack. Nonetheless, we see wine yeast being genetically modified so that the fermentation produces more resveratrol (Pretorius & Bauer, 2002).

There is an enormous diversity of strains of the yeast *Saccharomyces cerevisiae* (Pennisi, 2005) and the genetic properties of wine yeasts differ greatly from place to place. As a result the variation in commercial yeasts is also very great - accordingly there are as many wines as yeasts! There is still, however, a lot of research and optimisation to be done. The above-mentioned article by Pretorius and Bauer gives a good overview of the objectives of current research in this area. Basically it all comes down to improving the efficiency of the fermentation process and the subsequent steps, suppressing microbial decay, and above all increasing the health benefits and, of course, improving the flavour, colour and the bouquet of wine.

---

[37] www.falw.vu.nl/en/research/health-sciences/people/martijn-katan/index.asp

6.9. MANIPULATION OF THE GRAPES

Vines are classified under the *Vitis* genus. This genus has two subdivisions, namely *Euvitis* and *Muscadinia*. One single type, *Vitis vinifera,* has its origins in Europe, in the Caucasus, while in China there are more than 30 indigenous vines, and in North and Central America more than 30 types have been characterised. *V. vinifera* is the most cultivated species and approximately 5,000 cultivars of this species are grown on a commercial scale. The same goes for grape cultivars as for yeasts: there are as many wines as there are cultivars. In short, no two wines are the same! Initially new cultivars were obtained primarily by randomly selecting natural mutants with increased yield and/or better grapes for winemaking. In the second half of the 20th century this was done primarily via a more targeted selection of clones, but in 1994 the first field trials took place with genetically modified vines[40]. In the EU so far seven field trials with GM grape vines have been executed. A notable example with gene-modified fungal-resistant grape vines started in 1999 in two areas in Germany, the Palatinate (Pfalz) and Franconia (Franken). The field trials were planned to last for 10 years, and examined mainly the varieties Riesling and Chardonnay, for which it had not been possible previously to breed fungus-resistant vines. As result of fear and political pressure the trials were suspended at the beginning of 2005. In the USA the number of trials is now around sixty. A region-specific example was

[40] www.gmo-compass.org/eng/database/plants/73.grape_vine.html

reported at the beginning of 2009. Researchers at the University of Illinois at Urbana-Champaign created a genetically modified grape, called Improved Chancellor, with resistance to the herbicide 2,4-D (Textbox 6.3). Chile and South Africa are also very active in this field: South Africa for instance in the context of the "Grapevine Biotechnology Program" by the Institute of Wine Biotechnology in Stellenbosch and Stellenbosch University (*Stellenbosch University News*, 31 August 2006). Vines as well as yeasts are being genetically modified in this program and field trials started in 2006.

The subject of GM grapes is discussed extensively in a second review article by Pretorius (Vivier & Pretorius, 2002). The authors present an overview of those properties that are desirable in vines and that can be worked on at the present time with the help of recombinant DNA technology. First in line is increased resistance to infectious diseases caused by moulds, bacteria and viruses. This is to be expected given that grapes are among the most heavily sprayed of all crops, requiring an average of 12 applications in a season (DeFrancesco & Watanabe, 2008). Second is higher stress tolerance, i.e. to drought, oxidation damage, temperature and osmotic and other abiotic stresses. Finally, improved properties concerning quality, such as sugar content and colour.

In their paper DeFrancesco and Watanabe address the following question: "With the genome of the grapevine in hand, how likely are oenologists and winegrowers to resort to genetic engineering to tackle the problems facing viticulture?" They start by saying that the complete genome sequences of two grapevine cultivars now grace the public genome databases, bringing the latest sequencing technologies (see also Chapter 13) to bear on one of the oldest uses of biotechnology – winemaking. "The grapevine genome now joins the august group of completed plant genomes, being the first freshly fruit crop … to be sequenced." Found amid the sequence of the genome of the two cultivars are over 150 genes for aroma and flavour, three times the number found in other flowering plants. This knowledge should be very encouraging to grape geneticists and breeders who are considering the use of recombinant-DNA technology. However, the fear of a public outcry against GM grapes may continue to stop transgenesis of grapevines from taking hold. The authors end by quoting Marc Fuchs,

---

TEXTBOX 6.3.

GM Grapes Raise Hopes for Midwest Wine Industry.[38]

On 1 January 2009 H. Sterling Burnett reported in the Environment & Climate News on the genetic modification of Midwestern grapes. He wrote: "One of the most effective, widely used herbicides in the United States – known as 2,4-D – has a serious drawback: It devastates grapes. That makes it very difficult to raise grapes in the Midwest, because 2,4-D is widely used on popular staple food crops including corn and wheat, and it can harm grapes up to two miles away from its point of application. … In the aftermath of an accident spill of the pesticide, the United States Department of Agriculture (USDA) found a soil bacterium with a gene that allows it to break down 2,4-D. Building on these findings, in 2002 Robert Skirvin, a plant biologist at the University of Illinois, secured permission to use the gene and transfer it into the Chancellor grape. Skirvin and his colleagues used standard genetic engineering techniques to do so. … Of the eight Chancellor grape plants eventually developed through this process, three retained the herbicide resistance gene. Cuttings from the Improved Chancellor plants, along with a non-modified Chancellor used as a control, were sprayed with relatively high amounts of 2,4-D. The modified Chancellor grapes proved resistant to the herbicide. … Once the grapes have been found safe to eat, the research team will have to work with a grape grower to produce a wine using Improved Chancellor. Even then, environmental activists are likely to mount legal challenges to the distribution of the grapes, further delaying their introduction into the marketplace, experts say."

Bearing in mind the "loud outcry" concerning GM Grapevines of the Institute of Science in Society[39] we reckon that the latter statement will prove to be true. The outcry reads: "Clearly these new developments are crying out for GM labelling at the very least, and a clean sweep of the regulatory regimes would not come amiss."

---

[38] www.heartland.org/publications/environment%20climate/article/24364/GM_Grapes_Raise_Hopes_for_Midwest_Wine_Industry.html
[39] www.i-sis.org.uk/GMGrapevines_and_ToxicWines.php

molecular biologist at Cornell University since 2004 after working in viticulture for twenty years in France. He acknowledges grower resistance to GM wines and notes: "The wine industry is not very receptive to any innovation." However, over the past five years, he has detected a shift in mindset among the people with whom he interacts. "Most growers are convinced that science should move forward in case scientists have an interesting plant to offer the industry. They could then make the choice to use it or not," he says, but Fuchs is optimistic that the path to commercialisation is opening up. And so are we. In the above-mentioned article, Vivier and Pretorius also look at the obstacles that must be overcome if genetically improved vine cultivars are to be marketed. For example, in the area of:

- *science and technology;*
- *legislation;*
- *intellectual property and patents;*
- *politics and economics;*
- *marketing;*
- *tradition and culture;*
- *social acceptance.*

What is clear from these very informative review articles by Pretorius *et al.*, and also from other earlier articles from this group, is that the authors are in favour of using modern biotechnology in the wine industry. However, the articles also show that in 2002 there were still many obstacles on the road to commercial implementation. Their tone was clearly more optimistic

in an article published in 2005 (Pretorius & Hoj, 2005). According to them, the technological possibilities of modern biotechnology will cause a paradigm shift: the current process-oriented wine industry will become consumer-oriented, and thus market-oriented. This market demands wine that stimulates all our senses (except hearing, although ... think of the crystal clear sound of clinking wine glasses), that is beneficial for our health and that can be produced sustainably. Moreover, it wants wines that continue to be shrouded in the undefined mystique we associate with them (Bisson, Waterhouse, Ebeler, Walker, & Lapsley, 2002). For that, no two wines should ever be the same, that is our message.

6.10. WINEMAKERS RAISE THEIR GLASSES TO BIOTECHNOLOGY

This was the headline to an article about wine in the July 2006 newsletter of *LIS Consult,* a small Dutch consultancy office in the field of biotechnology. Readers can deduce from this headline that winemakers have something to celebrate as far as biotechnology is concerned. The same conclusion could be derived from the headline to an article in *AgraFood Biotech* of 13 February 2006: Wine industry gradually accepting GM. Not surprisingly, since both articles use the same sources[41, 42]. The LIS Consult article reads as follows:

"Winemaking is steeped in tradition. Connoisseurs place a great deal of value on the production conditions

[41] www.checkbiotech.org
[42] www.whybiotech.com

(soil, climate, aspect of the slopes, etc.), and the maturation and the development of taste and bouquet. From a technological point of view, much has changed in recent decades in the process of winemaking, but the cultivation of the grapes and the variety of grape being grown has continued virtually unaltered. Now, though, changes are afoot in that area. Until recently winegrowers, both in Europe and the US, were fierce opponents of the advent of biotechnology in grape cultivation. But in a lengthy article on the website of the Canadian Council for Biotechnology Information there is mention of a clear policy change among a number of leading organisations in the wine sector. As an example, *Winetech*, a professional association of South African winemakers, recently accepted a resolution expressing support for the introduction of biotechnology 'to help in the development of the tradition and science of winemaking'. The policy is aimed at 'promoting innovative research and dynamic science in a responsible and intelligent way'. In California the representatives of the *wine counties* have expressed their support for biotechnology because they expect it to play a key role in making grape cultivation sustainable in the long term. This, in spite of the fact that some counties have accepted non-GMO resolutions. Another sign that changes are afoot comes from wine countries around the world. Australia, Canada and various West European countries have started field trials. And according to C. Ford Runge, a professor and director at the University of Minnesota Center for International Food and Agricultural Policy, lab and greenhouse trials have also been reported in Chile, Eastern Europe and South Africa. Research is even underway in Italy. And in the Colmar region of France, small-scale field trials have already been in progress for some years. In addition to the development of grape cultivars that are resistant to a number of very common diseases, research is also being conducted into whether it is possible to grow 'healthier' grapes. In recent years much has been written about health-promoting substances particularly in red wine, e.g. sterols, polyphenols and antioxidants, which are micronutrients that may play a role in preventing cardiac disorders. The aim is to try to increase concentrations of these components using biotechnology, and to make wine even 'healthier'.

On top of all that, work is also being done on improving the yeast types that are used to make wine. For industry, the reduction in maturation time, which translates into cost savings, is what appeals most. As with many other reports on its website, the Council for Biotechnology Information also tries to present this area of application in a good light. Whether or not the consumer will accept all of this is obviously the million dollar question, because wine is for many consumers a hand-crafted product, where tradition plays a key role."

It is clear that authorisation committees such as the EFSA will continue to pour over this sensitive issue in great detail. All that remains is the question of when the first wine will be sold in which not only enzymes and yeast as the products of modern biotechnology are used in the production process, but where the grapes are also taken from a genetically modified plant.

6.11. IN CONCLUSION

The signs are clear: recombinant DNA technology is gradually gaining acceptance in the vineyard and winery. Stellenbosch in South Africa was way ahead of the game. Sadly for this region, Pretorius, who is so convinced of the many opportunities of modern biotechnology, has now found employment in a wine institute in Australia. A former colleague of his from Stellenbosch, Professor Hennie van Vuuren, also left Stellenbosch and has meantime become the director of the Wine Research Centre at the University of British Columbia in Vancouver. He was the person responsible for making and marketing a malolactic and an ethyl carbamate-free wine yeast using genetic modification (Pretorius, 6 August 2008, personal communication). They are now being used by several wineries in North America for commercial wine production.

We conclude this chapter with a free interpretation of a piece from the article by Pretorius & Hoj (2005).

The image of wine as a harmonious blend of nature, art and science has long been a source of tension between tradition and innovation. At the beginning of the 21st century this tension has increased as a result of the innumerable promising possibilities that modern biotechnology now offers for the genetic modification of grapes and yeast. The greatest challenge is to make the most of these possibilities without removing the charm, mystique and romance from grape cultivation and winemaking. An equally huge challenge is the great number of complex and interconnected obstacles, put there primarily by legislation and social reticence. These are currently blocking the commercial availability of transgenic grape cultivars, genetically modified yeasts and malolactic bacterial starter cultures. It goes without saying that a thorough investigation of the potential negative effects of new technologies is necessary, but if these hurdles cannot be removed within a reasonable timeframe, both the individual consumer and the whole international wine industry will be the loser. The previous sections have included examples of improved grape cultivars, yeast strains and bacterial starter cultures. Together these can make wine production more efficient and more cost-effective with a maximum profit in terms of product quality and a minimum consumption of raw ingredients and damage to the environment. In short, should we opt for a product that is better value for money in all aspects if we implement modern biotechnology, or should we stick to traditional ways on the basis of often irrational motives?

6.12. SOURCES

Baldwinson, J. (1975). *Plonk and Superplonk*. London, Coronet Books.

Bisson, L. F., Waterhouse, A. L., Ebeler, S. E., Walker, M. A., & Lapsley, J. T. (2002). The present and future of the international wine industry. *Nature, 418*(6898), 696-699.

DeFrancesco, L., & Watanabe, M. (2008). Vintage genetic engineering. *Nature Biotechnology, 26*(3), 261-263.

Pennisi, E. (2005). Wine yeast's surprising diversity. *Science, 309*(5733), 375-376.

Pretorius, I. S., & Bauer, F. F. (2002). Meeting the consumer challenge through genetically customized wine-yeast strains. *Trends in Biotechnology, 20*(10), 426-432.

Pretorius, I. S., & Hoj, P. B. (2005). Grape and wine biotechnology: Challenges, opportunities and potential benefits. *Australian Journal of Grape and Wine Research, 11*(2), 83-108.

Vivier, M. A., & Pretorius, I. S. (2002). Genetically tailored grapevines for the wine industry. *Trends in Biotechnology, 20*(11), 472-478.

"According to the law of Torah it is forbidden to let animals suffer."

In Judaism the word *Torah* is normally used to describe the first five books of the Hebrew Bible. These Five Books of Moses form the basis of the Jewish faith. Caring for animals is thus not only from this time as appears from this old Jewish saying. There are many traditions and rituals involved in the slaughter of cattle for human consumption, whereby concern for the animal's welfare is also a priority. For example, Muslims can only eat meat that is *halal*, in other words, meat that is slaughtered according to strict guidelines. One of the conditions is that the butcher should ensure that the animal is comfortable. The Jewish religion has similar guidelines (*kosher*). Meat is consequently a very traditional product, but not a traditional biotechnological product. Nevertheless, for decades meat has undergone processes involving the use of enzymes. These processes can thus justifiably be called biotechnological. In addition, in the production of some sausages there is a fermentation step that gives the sausage a slightly sour taste. This fermentation is sometimes helped along with the addition of bacteria cultures. As with bread, cheese and wine, which are really traditional biotech products, it is clear that modern biotechnology is making ever bigger inroads into cattle breeding and the meat industry. A number of topical examples are discussed in this chapter.

Chapter 7: Meat from the biotech vat

It will be a while yet before we find any meat on our plates that comes from genetically modified (also called transgenic or recombinant) animals. It is important, however, that we start thinking about this issue now, in the light of the many developments in this area. And just because it isn't currently happening, doesn't mean that modern biotechnology has not already made some inroads into the cattle rearing and meat processing industry. On the contrary, this is presently happening in the area of animal feed, not to mention the injection of hormones into animals. In both cases recombinant DNA technology is already playing a leading role. Enzymes in meat processing are a fact of life now, and it is likely that some of them come from genetically modified microorganisms. All these issues are discussed in turn in this chapter. Towards the end there are a few short sections on meat from cloned animals, other new developments and biotechnological meat substitutes.

7.2. ANIMAL FEED

Introduction

Commercial animal feed consists mainly of barley, rye and wheat grains. Corn and soya are also important components. The composition of animal feed is such that it facilitates the most efficient uptake of carbohydrates, proteins and fats by the cattle. The grain type and variety, the weather and in particular the ripeness of the grain, largely determine the available nutritional value which can vary greatly for chickens, pigs and cows. The addition of enzymes to animal feed can drastically increase the availability of the various nutrients, resulting in possible improvements in the efficiency of feed utilization and in positive environmental effects. Many of these enzymes are made using genetically modified microorganisms. There is an increasing likelihood that the most important components, in particular corn and soya, will also be products of modern biotechnology. As far as social acceptance is concerned, the use of these transgenic crops has huge consequences.

Milk substitutes

In animal husbandry in West Europe and North America it is common practice to separate the 'production mother' as soon as possible from her young. The sooner a suckling piglet or calf can be weaned off maternal milk, the better for the cattle breeder, because the mother can then quickly be fertilised again; the suckling of young animals has a contraceptive effect. Feed made from grain and vegetal proteins is not usually given to young piglets, because their stomachs and intestines are only able to tolerate milk. However, the addition of (recombinant) enzymes (see next page) facilitates the digestion of these vegetal proteins, enabling the pig farmer to give a cheaper blend of feed to the piglets. In some cases it is even possible, with the addition of enzymes, to completely replace the creamed-off milk, which is normally used as a substitute for maternal milk, without slowing down the growth of the piglets.

LOTS OF MILK SUBSTITUTES ARE AVAILABLE THESE DAYS THANKS TO BIOTECHNOLOGY

Transgenic crops

In the chapter on bread we saw how society's demand for a separation of the supply flows of genetically modified and conventional crops is becoming increasingly urgent. In order to achieve an effective separation of the two flows they must be processed strictly separately, and there must be proper methods of demonstrating the separation. Only then are labels reliable. Only then can the consumer make an informed choice and be sure of whether he/she is buying *recombinant-free* food or not.

Truly "recombinant-free" meat obviously requires that the animals from which it is sourced are fed with crops that also bear the label "recombinant-free". A still unanswered question is what that implies in terms of cost and whether the average consumer is prepared to pay that price.

Enzymes add nutritional value

Still another question is whether all animal feed additives should also be recombinant-free. Enzymes made using recombinant microorganisms play an important role in the production of animal feed, just as they do in cheese, bread and wine. These catalysts can ensure better digestion and take-up of the nutrients in the feed. As has been previously described, commercial feed mixes are blended to provide cattle with an optimum mix of starch (carbohydrates), proteins and fats. Maize is usually the most important source of energy in poultry diets. But if, for economic reasons, different grain types like barley, rye or wheat are used instead, productivity is considerably lower than expected from the nutrient content, because digestibility is poor. Furthermore, a similar alternative diet often causes sanitary and health problems such as sticky excrement and poor quality manure, with all the consequences thereof (infection, soiled eggs, etc.). The extent of the problems is closely related to the sort of grain, the specific cultivar, the climatic conditions under which the crop was grown, and, most of all, the time at which the grains were harvested (the ripeness). These problems can largely be prevented by adding suitable enzymes, which, for example, break down grain-cell walls, facilitating the uptake of nutrients by the animals. The addition of enzymes has been occurring on some scale for the last 20 years. Recombinant DNA technology is making it increasingly attractive in terms of cost, because microorganisms can be optimised by genetic modification to produce the required enzyme.

Adding enzymes helps the environment

The addition of carefully selected enzymes to animal feed can have a beneficial effect on the environment, as has been mentioned above. Better digestion of proteins, carbohydrates and fats, the main components of the feed, means less and better quality manure. There is also an advantage in terms of minerals, especially phosphorus, which is an essential element for life and development. Sixty to sixty-five percent of phosphorus in grains is present in the form of phytate, which animals with only one stomach (e.g. chicken and pigs) find difficult to digest. This has major consequences both for the feeding of these animals and for the environment. Phytate forms complexes with the ions of a number of important other elements such as iron and zinc, which consequently cannot be taken up efficiently from the feed by the animals. In order to ensure that the animals get enough phosphorus, extra inorganic phosphate is added to the feed, because this can be successfully taken up, even with iron and zinc as counter ions. Most of the phosphorus from animal feed is, however, released into the environment via the manure in the form of phytate and as such is one of the biggest environmental problems of intensive farming, i.e. exuberant algal growth suffocating lakes and coastal waters (eutrophication).

Various microorganisms and fungi in particular, make phytases. These are enzymes that break down phytate into inorganic phosphate, amongst other things. This latter can be taken up successfully by animals with one stomach. It seems only sensible, therefore, to add such enzymes to animal fodder. In fact this has been happening in recent years, mainly as a result of an EC ruling of May 2000. Since the end of the 1980s meat and bone meal are no longer permitted in ruminant feed, because of the risk of BSE (Mad Cow Disease). On 1 January 2001 an extended ban was introduced under pressure from consumers. As a result, these cheap sources of protein and phosphorus are no longer available to producers of pig and poultry feed.

In principle, then, the addition of phytases has a three-fold positive effect. Iron, zinc and other essential metals are better taken up; no extra phosphate is needed in the feed; and less phosphorus is released into the environment via the manure. Tests have substantiated the latter effect: up to 42% less phosphate in chicken manure and up to 35% less in pig manure has been observed.

LESS PHOSPHORUS IS RELEASED INTO THE ENVIRONMENT...
...AT NIGHT IT USED TO BE QUITE A SHOW!

Yet phytase is still only added to animal feed in a limited number of countries. This is partly due to its limited availability, since microorganisms don't make very large quantities of them. It is therefore only to be expected that some companies have tried to use copies of the phytase gene in better production organisms by genetically modifying them. DSM-Gist is one of the companies that succeeded in this venture, bringing a commercial product for use in animal feed onto the market more than ten years ago. Since then, there has been much activity in this market. An alliance between Novozymes and Roche Vitamins Ltd was set up in January 2001 to make enzymes for animal feed, and guaranteed these companies a strong position in this market. In 2003 DSM-Gist acquired this vitamin concern from Roche. To prevent DSM-Gist monopolising the area of enzymes for animal feed, the EU anti-trust legislation required the company to transfer the phytase production process to BASF, with whom they had a similar alliance. Today a long list of manufacturers and suppliers is available. A fact sheet on phytase has recently been published by Jacela *et al.* (Jacela *et al.*, 2010); this fact sheet can also be found on the website of the American Association of Swine Veterinarians[43].

Another interesting development in this area came when the Dutch company Mogen inserted the phytase gene in the rapeseed plant by modifying the genetic material. In 1997 this company set up a joint venture with DSM-Gist called Plantzyme to commercialise these developments. The phytase was made in the seed of the recombinant rapeseed plant. The advantage of this is that harvesting can take place in the usual way. In addition, the enzyme in the seed is apparently so well protected from negative influences, that it remains active for longer. As a result the seed

can be stored for several years. The best thing about it is that the rapeseed can be added without any further processing to the animal feed; there is no need to remove the phytase from the seed. Chicken and swine feed already contains a little rapeseed, so by replacing normal rapeseed with this transgenic rapeseed, there is no longer any need to add phytase. Figure 7.1 shows just how well the chicks thrive on it.

Shortly before the turn of the last century, it was expected that commercial introduction would soon follow. However, that has not yet happened. Probably in part because resistance to transgenic plants has not declined, and partly perhaps because DSM-Gist earned plenty from the recombinant microbial phytase, and partly too because the expression level of the enzyme in rapeseed is low. In the meantime, Mogen and Plantzyme are defunct.

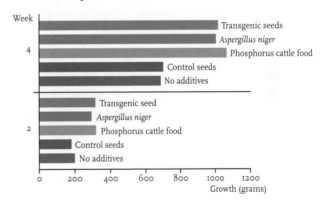

Figure 7.1. The effect of transgenic seed containing phytase on the growth of chicks over four weeks. For comparison the effect of feed with added Aspergillus niger (fungus) phytase, with inorganic phosphate, with conventional non-transgenic seed and finally feed without any of these additives (Koenderdam, 1997).

[43] http://www.aasv.org/shap/issues/v18n2/v18n2p90.html

"Approximately 80% of all beef production in the US is made possible thanks to the use of growth hormones."

In early 1996 the above statement was made in a Dutch agricultural newspaper by Philip M. Sheng, director of the US Meat Export Federation. What this quote shows is that the addition of hormones to cows is economically advantageous. The growth hormones in question are sex hormones like progesterone, testosterone, estradiol and synthetic hormones such as zeranol and trenbolone; in addition there is the recombinant bovine growth hormone *rBST*. By injecting rBST cows grow faster, milk yield increases, the meat is leaner, and the feed is used more efficiently, with less manure as a consequence. This obviously has a very commercial appeal to cattle farmers.

The journal *Genetic Engineering News* (GEN) has a column called "Point of View". In the edition of 15 January 1998 this column was entitled 'Public Education Still Needed on Biotech'. It was written by Isaac Rabino (1998), a professor of Biological and Health Sciences at the State University of New York. His article also maintained the following: that the complexity of the biotechnological discussion points is clearly illustrated by the production of rBST; and that action has been taken against the use of rBST by consumer groups, who feared that udder infection (mastitis) caused by increased milk production would lead to an increased use of antibiotics which would

result in a higher concentration of these in the milk, with all the consequences thereof. Others worried about further increases in overproduction and the consequent disappearance of the family farm. The use of rBST was therefore banned in Canada and the EU. In the EU this moratorium on the use of rBST and the other hormones came into force in 1988. There was also a consequent ban on the import of meat and milk products from cows injected with hormones.

Along with insulin and the swine diarrhea vaccine, rBST is one of the first commercial products of modern biotechnology and, as the previous paragraph shows, it has been controversial since the late '80s. By the mid '90s this discussion had flared up again in the EU. The daily and weekly papers began to express the view that it was no longer possible to uphold the European ban on rBST and on other growth hormones allowed in the US. A new global trade agreement, signed in

1995 by the EU and the US, included a statement declaring that this sort of hormone product, as well as genetically modified products, could only be banned on the basis of scientific arguments. In that same year an international committee of scientists, summoned by the EU, concluded that meat and dairy products from cows treated with hormones do not constitute a risk to consumer health, provided the hormones are administered to the cows under strictly controlled conditions. This paved the way for the lifting of the import ban on these products. At the end of 1995 the United States government then appealed directly to the WTO (World Trade Organization) to lift the ban on imports. Conversely, the EU did all it could to prevent the import ban from actually being removed. Earlier arguments, such as the supposedly excessively fast development of the skeleton with ensuing pain (growing pains), tumor formation, reduced fertility, increased stress and aggression among the treated animals, were all regurgitated.

In 1998 these American meat and dairy products were still not on the European market and the papers reported an escalation of the row into a full-scale war, in which talc and gelatin from hormone-induced cows and used in medicines, were thrown into the fray (see also Textbox 7.1). The financial stakes were also blown up out of all proportion. The discussion was not only to do with rBST and the other hormones permitted in the US. EU officials also dragged in the danger of BSE. This was referred to as the 'Mad Bureaucrat's Disease' of Brussels in American papers. As far as we know there is still no final agreement and the discussion continues about compensation rules and possible labelling of "normal" meat as hormone-free. It costs the EU more than 100 million dollars a year to maintain this recalcitrant attitude.

MAD BUREAUCRAT'S DISEASE

What is clearly undesirable is a repeat of the hypocritical situation we saw in the 1980s in Germany concerning insulin (see Section 1.5): American farmers being allowed to use hormones and therefore having an economic advantage on the European market with the free trade of their milk and meat products, because the European farmers are not being allowed to use these hormones. This is aside from all the other pros and cons in the matter of hormones being used to stimulate growth and milk yield in cows. In the US this feud has led to an increasing debate in society on such matters. One of the results of this is that on American supermarket shelves you can now find milk that is clearly labelled as 'guaranteed rBST free', both on the cap as well as on the shelf label.

The rBST affair reveals yet another distinct difference between the US and Europe, namely that of entrepreneurship. People in the USA are much quicker to turn a problem into an opportunity than those of us in Europe. The following example is a clear demonstration of that.

Gelatin is made from collagen. Collagen is the protein of the fibrous connective tissue of bone, cartilage and skin. It is the most prevalent protein in the higher vertebrates. Collagen, and therefore gelatin too, is obtained from animal carcasses. Gelatin is regularly used in food, but a secondary important application is in the pharmaceutical industry. It is an ingredient found in hard capsules, soft gels, plasma expanders, tablet binding agents and coatings, and vaccine stabilisers. Therefore, it is understandable that the US pharmaceutical industry reacted angrily when, as a result of the bovine hormone affair, the EU threatened to ban imports of drugs containing tallow or gelatin.

On 8 December 1999 there was an article in the AgraFood Biotech journal (Anonymous, 1999) claiming that the American firm FibroGen could successfully produce gelatin with genetically modified yeasts and plants (tobacco). According to the company, the main advantage was the safety of the product. Gelatin from animals carries the risk of being contaminated (with, for example, prions that cause Mad Cow Disease), or invoking the wrong immunological reactions in vaccinations against measles, mumps, scarlet fever, etc. The new process enables FibroGen to make 'customised' safe gelatins for specific applications. The company was extraordinarily quick to capitalise on the escalating rBST affair. When we looked at their website[44] again in 2010, we noticed that they now offer rDNA-yeast gelatins that are similar to human gelatin and that are non-immunogenic, as demonstrated by research in collaboration with groups that specialise in gelatin allergy.

In 2003 the biotech company Monsanto lodged a complaint against the dairy company Oakhurst Dairy from Portland, Maine in the US, because of a misleading claim. According to Monsanto the advertising done by this family company gave the impression that milk that didn't come from rBST cows was safer than rBST milk, while scientific studies demonstrated that this was not the case. What is remarkable is that the biotech company waited until this moment to take action, when the dairy company had been making rBST-free claims for years. Stanley Bennett, the director of Oakhurst Dairy[45], maintained that the company had been producing rBST-free milk for five years because consumers liked it: "We are in the business of marketing milk, not Monsanto's drugs."

Five years later, in 2008, the discussion flared up again in the US, while in the EU it hasn't been on

[44] www.fibrogen.com/collagen_gelatin
[45] www.oakhurstdairy.com/about

the news agenda for years. There was a heated discussion about labelling, supermarkets removed rBST products from the shelves and major dairy companies stopped buying and processing rBST milk. A report published in mid-2009 spoke about an overreaction and a storm in a teacup: only one in ten consumers is worried enough to want to change purchasing and consumption behaviour. Half of the "worriers" already buy alternative products.

DID YOU KNOW THAT PINEAPPLE IS USED TO TENDERISE MEAT?!

7.4. MEAT PROCESSING

Meat can be tenderised naturally by storing it for about ten days at 2° C. This slow process of converting muscle into meat with endogenous proteases (protein-digesting enzymes) and collagenases (endopeptidases) ensures tender meat; the downside is moisture loss and shrinkage. Research has been under way since 1940 to see if this "hanging" process ("aging" in game) can be improved by using exogenous enzymes. On a commercial scale proteases from plants, for example papain from papaya and bromelain

from pineapple, are now used to tenderise meat. It is simply a matter of time before enzymes made with recombinant microorganisms make their entry in this area too. Yet modern biotechnology already has its "foot in the door" in meat production.

It is extremely important to maximise the profit on commercial products of meat processing. To this end methods are being developed to restructure low-value pieces, scraps and juices to make them look tastier and more appetising and to enhance the flavour and texture, to raise the market value. These processes usually consist of cutting the meat into small pieces, then shaping and binding. For some years now Novozymes, the Danish enzyme manufacturer mentioned earlier, has been selling a mixture of specific proteolytic enzymes that dramatically improves this process. The enzyme transglutaminase can also make a very positive contribution in this area. This enzyme binds proteins, peptides and amino acids together, with the result that the texture is better, lysin - an essential amino acid - is better protected against chemical conversions, fats and fat-soluble substances are locked in, heat and water-resistant films are formed, heat treatments for jelly formation are unnecessary, elasticity and water-binding capacity are enhanced, solubility and functional properties change for the better, and the nutrition value increases because various proteins, whose composition is complementary to limiting essential amino acids, are bound to each other. In short, a whole gamut of positive effects may be achieved.

Until recently the only commercially available enzyme with this mechanism was transglutaminase isolated from the livers of Guinea pigs. The scarcity, and the difficult and laborious method of obtaining the enzyme in a workable form, made it extremely expensive - too expensive for use in industrial meat processing. New sources are now available. For instance, transglutaminases are found in microorganisms such as *Streptoverticillium* and *Streptomyces* strains. Several years ago the Japanese company Ajinomoto brought a transglutaminase from a microbial source onto the market. Its use in food production is, however, banned in EU countries. These microbial sources have highlighted cheaper production methods, especially if recombinant DNA technology is used, which is only a matter of time. This will open the way to efficient and profitable meat processing, and much more. Table 7.1 shows a list of a great many other interesting possibilities for food processing.

Table 7.1. Summary of application possibilities of the microbial enzyme transglutaminase in food processing.

Source	Product	Effect
Meat	Hamburgers, meatballs, dumplings, shaomai Tinned meat Frozen meat Compressed meat	Better elasticity, texture, taste and aroma. Good texture and appearance. Improved texture and lower costs. Restructuring of meat
Fish	Fish pie	Improved texture and appearance
Krill	Krill pie	Improved texture
Collagen	Shark fin imitation	Imitation of tasty food
Grain	Baked food	Improved texture with more volume
Soya beans	Mapuo-Doufu Baked Tofu (Aburaage) Tofu	Improved shelf life. Improved texture Improved shelf life
Fruit and vegetables	Celery	Food preservation
Casein	Stimulators of mineral absorption Cross-linked proteins	Improved mineral absorption in the intestine. Reduced allergen reactions
Gelatin	Sweet food	Food low in calories and good texture, form and elasticity
Fat, oil and proteins	Hard fats	Lard substitute with good taste, texture and aroma
Vegetable proteins	Protein powders	Gel formation with good texture and taste
Herbs	Herbs	Improved taste and aroma

7.5. CLONED MEAT

In Chapters 1 and 2 genetically modified, so-called transgenic animals, and their clones, were introduced. We will come back to this topic again in Chapter 14. The number of transgenic and cloned animals is still relatively small, but the number is rapidly rising. Even in a country like New Zealand, with campaigns against transgenic animals (see Section 2.2), the authorities granted permission in April 2010 to the country's leading agricultural research company to continue its work on genetically modified livestock[46]. This decision means the animals, which include a herd of 100 genetically modified (GM) cows, can be returned to an active breeding program. Not everyone is happy with this decision. "We are appalled," said Claire Bleakley of GE free NZ[47]. "They now have carte blanche to produce any number of GM animals with no way to properly assess the potential danger to health and the environment, and the controls are no stricter than those for previous decisions."

According to Jonathan Cowie (2000) there will be a major global food crisis in the middle of this century. He bases this conclusion on the report *"GM crops: The Social and Ethical Issues"* of 1998, which was drawn up by a working group (of which Cowie was a member) of the English Society of Biology and six affiliated associations (whose specialist interests range from agricultural production to ecological conservation). Such a food crisis will mean the inevitable use of everything that is edible. Sooner or later we will be confronted with the question as to whether meat from cloned animals is suitable for consumption. In fact we have already reached that point. A cattle company in Canada already applied in 2003 to the authorities to be allowed to sell meat from cloned cattle for human consumption.

A more recent development began on 28 December 2006 when the Centre for Veterinary Medicine (CVM) of the FDA published a three-part discussion document entitled: *A Risk-Based Approach to Evaluate Animal Clones and Their Progeny – Draft.* It consisted of a draft risk analysis, a proposal for a risk management plan and draft guidelines for the industry. Before the draft risk analysis was published, it had already been looked at by independent scientific experts, who agreed with the methods the CVM used to evaluate the data, and therefore supported the conclusions.

According to the CVM meat and milk from cloned cows, pigs and goats are just as safe to eat as conventional food and thus require no special labelling. The American Biotechnology Industry Organization (BIO) backed the CVM in this matter and said in a press release that the results are consistent with numerous studies already showing that food from animal clones and their offspring is safe. The CVM invited with the 2006 draft the American citizens to give their opinion on the document. BIO had this to say:

"While there are currently no products from cloned animals and their offspring in the market, the publication of the FDA's draft risk assessment

[46] www.abc.net.au/science/articles/2010/04/15/2873674.htm
[47] www.gefree.org.nz

will begin an essential public discussion on the technology and how it can be successfully used by farmers and ranchers."

The American Meat Institute responded by issuing a warning that the FDA should be extremely cautious about allowing such animal clone products on the market because many consumers have difficulty accepting this technology. They urge the government not just to uphold the safety of these products in the political arena, but also to make it easier for the consumer to obtain a better understanding of what cloning actually entails, so that the *overall* confidence of the consumer in food provision is maintained.

CONSUMERS HAVE DIFFICULTIES ACCEPTING BIOTECH FOOD

AND WHY DON'T YOU TRUST YOUR BUTCHER ANYMORE?

Proof that the meat institute's comment is not entirely wide of the mark appeared in an article in the *Washington Post* of 10 February 2007 (Anonymous, 2007). The headline runs as follows: "Frankenfood? Not quite."

The article quotes surveys showing that the average American is very concerned about the introduction of meat and milk from clones in the supermarket, and that members of Congress are preparing legislation requiring food from clones and their offspring to be labelled as such. According to one of their reporters there is also a threat of a new controversy about whether food manufacturers can label their products from the offspring of cloned animals as 'organic', if the farmers concerned comply with the requisite federal criteria. It is further stated that the debate about cloned food has hotted-up and that in 2008 there might well be a battle between the cloning industry, the anti-cloning supporters, the FDA and Congress. The article emphasises that the public should not overestimate the differences between cloning cattle to improve the breed, and what current-day cattle breeders are now doing to improve their herds. In conclusion, it is remarked that the facts are less threatening than many Americans believe.

On 15 January 2008 the CVM-FDA published a three-part final report of almost 1000 pages, the main conclusion of which was: Milk and meat from healthy clones and their offspring are just as safe for consumption as that from normal cows, pigs and goats. The scientific committee of the EFSA simultaneously brought out a draft opinion in support of this conclusion. The European Group on Ethics in Science and New Technologies (EGE[48]) took a fiercely opposing point of view, saying that the cloning of animals for food production is not justified for ethical and welfare reasons. And so began the commotion about

[48] ec.europa.eu/european_group_ethics/index_en.htm

this in the EU. In mid-2009 a complete ban on cloned meat in the EU seemed imminent. Even the then newly appointed MEPs seemed to have similar views to the EGE about it. A year later on 7 July, the day of the final revision of this chapter, the New York Times headed an article with: Europe Seeks to Ban Food from Clones. James Kanter, the reporter, opened with: "The European Parliament asked on Wednesday for a ban on the sale of foods from cloned animals and their offspring, the latest sign of deepening concern in the European Union about the safety and ethics of new food technologies."

What is clear to us is that not everything in the area of modern biotechnology is accepted unquestioningly by society, neither in the EU nor in the US. It might once have been the case for the US public, but we now see the American government itself encouraging a serious public debate about this thorny issue - something that is not really happening yet here in the Netherlands or anywhere else. Times change.

7.6. NEW DEVELOPMENTS

"Healthy eating is about as feasible as a healthy lamppost."

On Monday 25 June 2001 the Dutch public debate "Biotechnology and Food" began with the opening event "Food and Genes" (see also Chapter 8). At some point during this day there was an appearance by Midas Dekkers. His appearance was announced in the program booklet with the above quote. The booklet described him as follows:

"One biologist practises his profession with a pair of binoculars, the other with test tubes. Midas Dekkers (1946) does it with a typewriter, microphone or camera. His interest in animals grew from his time studying in Amsterdam, his interest in people was awakened in his parents' cafe. Midas Dekkers writes mostly about the common ground between both. Since this is rather remarkable, it is often rather amusing. Before you know it, you've really learned something."

HEALTHY EATING IS ABOUT AS FEASIBLE AS A HEALTHY LAMPPOST

What we remember are his statements about meat. He noticed that most people are very opposed to the thought that a piece of steak, or meat in general, could be made entirely within a factory. Remarkable, he says, because practically all our food is made for the most part in a factory. Now, however, industry is working on a recipe for minced meat from a lab.

Test-tube steak

"Researchers in the Netherlands have created what was described as soggy pork and are now investigating ways to improve the muscle tissue in the hope that people will one day want to eat it. No one has yet tasted the product, but it is believed the artificial meat could be on sale within five years."

These statements are made by Nick Britten in the Telegraph of 29 November 2009[49]. Four years earlier the headline to a full-page article by Annemarie Eek in a Dutch biology journal (Eek, 2005) had a more cautious message: "It sounds like science fiction, but Dutch researchers are cultivating meat from pig stem cells. Due to practical problems it will still be a while before cultivated meat appears in our supermarket." She wrote that the Netherlands was the first country to systematically conduct research into the development of cultivated meat. The idea is that stem cells (see also Chapter 14) multiply and then differentiate into muscle cells, which then fuse into muscle fibres. These fibres would finally form a real piece of meat together with connective tissue and fat cells. There's still a lot of research to be done, however, before this stage is reached. The expectations then were that it would be possible to make a sort of minced meat from the fibres

in about six years. But cultivating a real piece of meat would take at least twice as long. Sebastiaan Donders nicely illustrated this research process in the article and Figure 7.2 is a loose interpretation of it. In the article a few other curious studies in this area are also mentioned (Textbox 7.2).

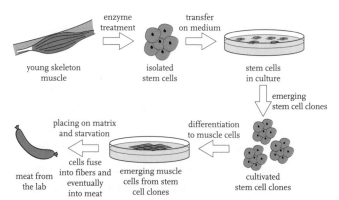

Figure 7.2. Steps in the development of meat from the lab, adapted from Eek (2005).

Vegetarian groups and animal rights campaigners see no ethical objection if meat was not a piece of a dead animal, says Nick Britten in the Telegraph article. Meat produced in the lab could also reduce greenhouse gas emissions associated with real animals, according to him. However, the Vegetarian Society said: "The big question is how could you guarantee you were eating artificial flesh rather than flesh from an animal that had been slaughtered. It would be very difficult to label and identify in a way that people would trust." A week earlier Prince Charles, a fierce opponent of GM food, warned that people were creating problems by treating food as an easy commodity

[49] www.telegraph.co.uk/science/science-news/6680989/Meat-grown-in-laboratory-in-world-first.html

Back in 1912 the Nobel Prize winner and surgeon Alexis Carrel succeeded in keeping a piece of heart muscle from a chicken embryo alive. Every two days the tiny muscle received fresh food and a clean bottle. The story goes that Carrel sang 'Happy Birthday' to his little muscle every year and that it wasn't until 32 years later that the piece of meat died, along with its carer.

In West Australia artists and tissue engineers collaborated on cultivating meat. It didn't work with stem cells from sheep, but pieces of frog muscle grew a tiny bit. The minute pieces of frogs' legs, in Calvados sauce, surrounded by a host of live frogs, were presented at a bio-art exhibition in Nantes in March 2003.

A more serious attempt to develop cultivated meat was made by tissue engineers at Touro College in New York at the behest of NASA. In order to be able to provide meat for astronauts in space, they cultivated goldfish tissue in a Petri dish. Unfortunately, the meat only grew on serum from calf foetuses, which meant that still more animals were needed for its production.

rather than a precious gift from nature. The real problem is in our opinion the answer to the question "how are we going to feed the world in 2050 in a sustainable way?"

7.7. BIOTECHNOLOGICAL MEAT SUBSTITUTES

There is an urgent need for new sources of protein as an alternative to meat. This is because of the growing world population and increased affluence in countries such as China and India. Back in the 1950s oil companies, and not the food industry, had already begun producing microbial protein as a meat substitute, the so-called *single cell protein*. They used by-products as a substrate for cultivating certain types of microorganisms. British Petroleum (BP) brought Tropina onto the market, a protein originating from a yeast that was grown on alkanes. BP first studied Tropina for twelve years for toxicity and carcinogenic properties, but found no proof of harmful effects. But consumers still doubted the safety of Tropina, fearing that it contained aromatic hydrocarbons. Opposition in Japan even led to a complete ban on edible proteins produced by the petrochemical industry. Governments around Europe demanded more research. Further studies showed that the product was not carcinogenic. On these grounds Tropina was permitted, albeit in restricted quantities and only for export! However, when the price of oil began to rise, the substrate on which the yeasts were cultivated became quite expensive. BP therefore decided to stop production of Tropina, because it could no longer compete with soya protein (Israelidis, 1988). The British company ICI ran into the same problems. It introduced Pruteen, a protein made from bacteria grown on methanol as feed. The raw protein content of Pruteen was 72% and it was brought onto the market as a high-value protein. In other words, it had a well-balanced amino acid composition. ICI was building a Pruteen factory with

60,000 ton capacity per year when the company was confronted with the oil price rise. Despite successful engineering, it turned out that Pruteen too could no longer compete with proteins from soya and fish.

A microbial protein that enjoyed more success is mycoprotein. It was produced with a *Fusarium* fungus and processed into Quorn products. In 1986 Quorn was launched on the British market as a meat substitute. These products satisfy a number of important consumer needs. For example, they are healthy, easy to prepare and have the same flavour and texture as normal food. Quorn is now seen as more of a meat substitute than, for example, soya protein.

A completely different and interesting group of organisms that can be used as a source of protein are insects. In the Western world this is virtually unheard of as a food group, but it is becoming increasingly important. Edible insects (more than 1,380 types) have long been accepted as a nutritious source of proteins, vitamins and energy in many non-Western countries. Not all varieties are suitable, however, for large-scale breeding, because they are susceptible to disease. Modern biotechnology does, however, offer an alternative in the form of insect cell cultures, whether or not they are infected with a (recombinant) virus. Insect cells can be grown in bioreactors on a large scale under controlled, closed conditions. There are also several possible ways of purposely and rationally changing the composition of insect cells, thereby changing the nutritional value. Whether this is feasible on a technological level and whether the nutritional value of insect cells is comparable with

that of whole insects still needs to be investigated. In addition to these technological aspects it is vital to find out whether consumers will accept food made from insect cells and under what conditions, and whether this can be influenced by a suitable marketing strategy (Verkerk, Tramper, Van Trijp, & Martens, 2007).

BIOTECHNOLOGICAL MEAT SUBSTITUTES

ONE 'BIG HOPPER' MENU PLEASE!

7.8. IN CONCLUSION: HAPPY MEAT!

In 2006 the Dutch VPRO science programme NWTV made an online appeal inviting people to come up with a new name for cultivated meat[50], meat-like material thus that is made from muscle cells grown in bioreactors. Three of the more than 200 entries were rewarded by the jury with an internet radio. The

[50] http://noorderlicht.vpro.nl/artikelen/28570228/

winners were *La Box* from Joop de Meij (a vegetarian), Ewart Kuijk with *Kreas* (Greek for meat) and Dafne Westerhof from pork paradise The Promised PigLand with *Happy Meat:*

"Then you know you're going to eat a tasty meal and no animal had to suffer for it."

Isn't that what we all want?

Note added in proof:
The British food authority FSA reported in August 2010 that meat from a cloned bull had ended up in the food chain. Moreover milk of offspring of a cloned cow has also reached consumers.

7.9. SOURCES

Anonymous. (1999, 8 December). Gelatin to be manufactured recombinantly. *AgraFood Biotech,* p. 28.

Anonymous. (2007, 10 February). Frankenfood? Not quite. *Washington Post.*

Cowie, J. (2000). Genetic modification and the meat market. *Nature, 404*(6781), 921-922.

Eek, A. (2005, 30 September). Biefstuk uit het buisje. *Bionieuws.*

Israelidis, C. J. (1988). Nutrition – single cell protein, twenty years later, *Biopolitics: Proceedings First Biointernational Conference* (Vol. 1).

Jacela, J., DeRouchey, J., Tokach, M., Goodband, R., Nelssen, J., Renter, D., & Dritz, S. (2010). Feed additives for swine: Fact sheets–high dietary levels of copper and zinc for young pigs, and phytase. *Journal of Swine Health and Production, 18*(2), 87-91.

Koenderdam, I. (1997, 18 January). Eerste product Mogen nadert de markt. *Chemisch Weekblad,* p. 1.

Rabino, I. (1998, 15 January). Public Education Still Needed on Biotech. *Genetic Engineering News.*

Verkerk, M., Tramper, J., Van Trijp, J., & Martens, D. (2007). Insect cells for human food. *Biotechnology Advances, 25*(2), 198-202.

"My concern is if we don't have a broadly educated public … charlatans out there will be able to play on public fears."

This is a quote from Donna Shalala, a professor of Political Sciences at the University of Miami, where she has also held the post of *President* since 2001. Donna Shalala is considered to be one of America's best leaders[51] and is not afraid of controversy, as demonstrated by the above forthright remark she made during a lecture on gene technology for scientists. Unfortunately her fears are not unfounded and the concerns many people have about genetically modified (transgenic) food crops are largely based on misleading information. The objectionable term *Frankenfood*, used by opponents to describe food made from what they call genetically manipulated plants, conjures up negative associations. The size of the rift between biotechnologists and the anti-GM food lobby and the extent of the unwillingness of either to reach out to the other is plain to see.

In this chapter we will look at a number of topics related to genetically modified food. The first section will deal with our food production and more specifically "GM foods". Secondly we will look at whether the concerns about GM foods are justified. In the third section we will consider whether GM foods are harmful to our health. After this section we will see that it is not just consumers but also farmers who are concerned about the production of GM foods. In the fifth section we will ask who is telling the truth in the debate about Frankenfood. In the last section of this chapter we take a look at the future of these crops. The idea behind this chapter is to present readers with enough information so that they can form their own considered opinion on this subject.

51 www.usnews.com/usnews/news/articles/051022/22shalala.htm

The science on which gene technology is based is not exactly straightforward, and yet it should still be possible for laymen to make reasonable choices based on good, objective information, each according to their own insight, beliefs or principles, and not on fear. In the media a lot of time and space has always been devoted to this subject, but consistent information of any real scope aimed at the general public only appeared in the Netherlands in 2001, when the government set up the temporary Terlouw Committee.

Jan Terlouw, the chairman of the committee, is a physicist, writer and politician. Even after his retirement in 1996 he remained very active and is still a great advocate of human and animal rights. This Biotechnology and Food Committee, also called the Food and Genes Committee, carried out a survey of what the public in the Netherlands thought about the use of biotechnology. It also attempted to make accurate information accessible to a broader audience. In its report[52] (2002, in Dutch) the committee maintained that there was no question about the safety of food prepared using modern biotechnology and that the majority of the public thought this technology should be allowed to develop further. The report made several important recommendations including better information, freedom of choice for the consumer and the setting up of an independent Food Authority.

Shortly thereafter, on 10 July 2002, the Food and Consumer Product Safety Authority (VWA) was founded by the government with the following mission[53]: "The Food and Consumer Product Safety Authority works on safe and healthy food, safe products and healthy animals. To this end, the VWA looks at the risks, evaluates them, communicates about them and makes them manageable within society." An ambitious, heavy and difficult task, which from a structural point of view requires a lot of money. By 2008 budget cuts had already prevented the VWA from being able to function as required. On 20 June 2008 a Belgian agricultural expert wrote a very plain-speaking report about it, declaring that the VWA was not functioning properly and was being patronised by the Ministry of Agriculture. Janneke Snijder, Member of the Dutch House of Representatives for the liberal party, concluded that the House of Representatives and the Agriculture Minister were heading for a showdown about the humiliatingly bad organisational structure of the VWA. The majority of the House wanted the whole situation to be sorted out, before merging the VWA, according to the most recent plan, with the General Inspectorate and the Plant Disease Service (these services are both part of the Ministry of Agriculture). The Minister states conversely that the merger should just go ahead because "it will be years" before the situation at the VWA improves. It seems to us there is little hope therefore of the VWA being able to complete its mission in the near future and to play a leading role in removing irrational fears about genetically modified

---

[52] www.voedingscentrum.nl/resources2008/eindrapport_terlouwpdf.pdf
[53] www.vwa.nl

food; a situation quite typical for the EU as a whole.

At the time of the Terlouw Committee, in 2001, a comprehensive report appeared on the Internet about this controversial topic (*A Report on Genetically Engineered Crops*). The author, Charles M. Rader, is himself an outsider in this area. Using a great many verifiable facts he made a careful analysis of the various aspects and put it online in layman's terms. We gratefully made use of this report to put together the first version of this chapter several years ago. For this final version we have again, though to a lesser extent, used the January 2008 revised version of the Rader report. We have now, for the revision, especially used information from a dozen or so publications that appeared recently in leading scientific journals. However, it is still worth visiting Rader's website[54].

PRINCE CHARLES: 'MANIPULATION OF GENETIC MATERIAL SHOULD ONLY BE THE WORK OF GOD'

I CAN'T FIND THE TERMS 'BIOTECHNOLOGY OR GM' IN ANY OLD RELIGIOUS BOOK!

8.2. JUSTIFIED FEARS?

"Ye shall keep my statutes. Thou shalt not let thy cattle gender with a diverse kind: thou shalt not sow thy field with mingled seed: neither shall a garment mingled of linen and woollen come upon thee."

Leviticus 19:19, King James version.

Some people, among them the UK's Prince Charles, believe that the manipulation of genetic material should only be the work of God or Mother Nature. We respect this point of view, but we don't think it can be the basis for an argument to ban gene technology, without taking into consideration other, far-reaching, consequences. Leviticus 19:19 forbids cross-fertilisation, i.e. sexual reproduction of two different plants or animals (species or cultivars). At this point it is worth mentioning that natural cross-fertilisation between different species is not at all rare, for example, cross-fertilisation in nature between different types of birds is a known occurrence. If improvements made possible by genetic modification using recombinant DNA technology fall under this ban, then strictly speaking many other, long-used breeding techniques must also be excluded. Obviously it should be possible for consumers to decide whether or not they want to buy such 'cross-hybridised' products. Vegetarians must be able to choose food that is free of animal genes and Muslims and Jews should be able to buy food free of pork and other forbidden meat products, even in terms of genes.

In order to make these choices accurate labelling is vital, as also recommended in the Terlouw Committee's report. Since this committee was set up, far-reaching

[54] members.tripod.com/c_rader0/gemod.htm

legislation, discussed in Chapters 2 and 5, has been introduced both at a national and EU level. For many people, though, there is still the question of how certain we can be that genetically modified food is not harmful to our health, in the short or long term. There is also some doubt as to whether transgenic crops will solve the world's food shortage problem and will result in healthier and more complete food, and whether they are damaging for the environment in the long term.

The precautionary principle (O'Riordan & Cameron, 1994) is central to environmental policy (see Textbox 3.3 in Chapter 3). If we approach food safety from the precautionary principle, then all significantly modified crops must be carefully evaluated. This would apply not just to crops that have been modified with recombinant DNA technology. In contrast to the more traditional breeders, molecular biologists that genetically modify plants using recombinant DNA technology have a pretty good idea what they are doing. In other words, genetic modification is much more of a targeted process than traditional breeding, which relies largely upon chance. Unexpected changes can still occur with genetic modification, but from a scientific perspective there is no reason why GM foods should be more thoroughly tested than other new food, except with a view to removing public concern.

The following example demonstrates the need for, and efficacy of, testing. Soya is a rich but low-quality source of protein compared to animal proteins because it contains very little methionine - one of the essential amino acids. Researchers have therefore genetically modified the soya plant with a gene coding for a protein rich in methionine. The gene came from a Brazilian nut to which some people are allergic. Tests have shown that the enriched protein from the transgenic soya also causes allergic reactions in these people. The transferred gene therefore seems to code for the protein that causes the allergy. For this reason the project was suspended long before there was any talk of marketing or consumption. This example therefore shows why testing is necessary, that it happens and that it safeguards the consumer. Ironically, opponents of gene technology have a different take on this example: "This just goes to show what can go wrong." What is clear is that (products of) transgenic crops should be and are more stringently tested and regulated than 'normal' crops, especially in the EU (S. H. Morris, 2007). To date the safeguards seem to be more than adequate: there are no indications that anyone has been exposed to unsafe GM foods in the more than ten years during which they have been consumed on a large scale, especially in the US. That's not something you can always say about "normal" food.

8.3. ARE GM FOODS HARMFUL TO HEALTH?

As mentioned above, in 2002 the Terlouw Committee maintained that there should be no doubt about the safety of food prepared using modern biotechnology. According to information from 2002 (*20 questions on genetically modified (GM) Foods*)[55], from the

[55] www.who.int/foodsafety/publications/biotech/20questions/en/index.html

In 2003, the Codex Alimentarius Committee, under the auspices of the FAO and WHO, adopted guidelines to harmonise the premarket process of risk assessment of genetically modified plants intended for the global market. These guidelines are intended to help individual countries draw up consistent legislation that provides for a strong food safety evaluation process, while avoiding trade barriers. Every new gene crop has to undergo a premarket safety assessment to evaluate intentional and unintentional changes for any adverse effects on human health, which are caused by introducing recombinant DNA (genes). The aim is to identify hazards and, if any are found, to demand a risk assessment, and if necessary a risk management strategy, for example, non-approval, approval but with labelling and/ or monitoring required, or approval without restriction.

The process is based on science and requires the use of methods and criteria which are proven to be predictive. New methods must be validated and it must be shown that they improve the risk assessment.

The framework for evaluating potential safety risks

requires detailed characteristics of:

- *the genetically modified plant and its use as food;*
- *the origin of the gene;*
- *the inserted DNA and flanking DNA at the place of insertion in the genome;*
- *the substances expressed (for example, proteins and any new metabolic products that are the result of the new gene product);*
- *the possible toxicity and anti-nutritional properties of new proteins or metabolic products;*
- *the protein expressed by the new gene compared to that which is known to cause Coeliac's, if the DNA comes from wheat, barley, rye, oat or related grains;*
- *the protein expressed by the new gene with regard to possible allergy;*
- *key endogenic nutrients and anti-nutrients including toxins and allergens for possible increases for specific host plants (the DNA recipients).*

Some steps of the risk assessment require a scientific assessment of existing information; others require experiments, in which case validation of the analysis, sensitivity and verifiable documentation is required.

World Health Organization (WHO), all genetically modified food products that are now available on the international market, have undergone a risk analysis by national authorities in compliance with the Codex guidelines (Textbox 8.1). These analyses have shown no proof of any risk to consumer health.

WHO indicates that every GM food must be tested separately, and subjected to individual safety tests, because different genes are inserted in different ways in different GM crops. It is therefore impossible to make a generally applicable declaration concerning the safety of GM foods. In general terms, the safety assessment of GM foods should, according to WHO, investigate:

- *toxicity;*
- *allergenicity;*
- *specific components thought to have nutritional or toxic properties;*
- *stability of the inserted gene;*
- *nutritional effects associated with the genetic modification; and*
- *any unintended effects which could result from the gene insertion.*

A major problem when conducting safety assessments on GM foods is the use of the concept of *substantial equivalence*. This concept is based on the principle that "if a new food is found to be substantially equivalent in composition and nutritional characteristics to an existing food, it can be regarded as being as safe as the conventional food." The problem is obviously the definition of the word 'equivalent'. Kuiper *et al.* (2002) state that this concept is not a safety assessment per se, but does enable us to identify possible differences between existing food and a new product. Extra attention can then be focused on the differences in terms of the toxicological aspects. It is more of a starting point than an endpoint.

Thus, safety assessments are conducted on all the GM foods brought onto the market. To date there has been no evidence that eating such food carries any risk for our health. Yet Domingo (2007) in his critical literature review of toxicity studies on genetically modified plants concludes with the question: "Where is the scientific evidence showing that GM plants/foods are toxicologically safe as assumed by the biotechnology companies involved in commercial GM foods?" This question arises from his analysis of the surprisingly limited number of scientific articles concerning human and animal toxicological and health risk studies on GM crops, such as potatoes, corn, soya beans and rice. In addition, it appears that, of the few scientific publications available, most do not originate from companies that grow transgenic crops or make products from them. Domingo wonders how that is possible, given that the debate on the safety of GM foods is causing such controversy. His overview shows that these are mainly short-term studies, in which nutritional values rather than toxicological aspects are examined. He concludes that long-term studies are urgently required, where attention is devoted primarily to (1) people with abnormal digestion as a result of chronic gastrointestinal disease, and (2) undesirable DNA transfer into mucosa and intestinal flora.

According to the European Food Safety Authority (Anonymous, 2008a; EFSA, 2008) (see also Chapters 2 & 3), 90-day food trials with rodents, mainly rats, are generally sufficient to show that GM foods are safe, provided that they are conducted in compliance with international guidelines. This was the conclusion of the EFSA in a report published at the beginning of March 2008 in the scientific journal *Food and Chemical Toxicology* (van Haver *et al.*, 2008). It was stated, however, that these trials are not suitable for identifying potential allergens. The report calls for a more uniform approach to food testing and the use of new profiling technologies. It rejects monitoring following a product's entry on the market; in any case it should in no circumstances be a substitute for a solid preliminary risk analysis.

The first 2008 issue of *Nature Biotechnology* (Goodman *et al.*, 2008) contains an article by seven allergy experts, entitled "Allergenicity assessment of genetically modified crops – what makes sense?" The seven maintain the following:

- *GM crops offer major opportunities for improving food quality, increasing harvest yields and reducing dependence on some pesticides.*
- *Before they are brought onto the market, they should first be subject to a very stringent safety test, as well as a detailed analysis of the allergenic risks.*
- *There is (still) no documented evidence that the current approved commercially grown GM crops caused allergic reactions as a result of an allergenic protein, coded by the gene inserted by genetic modification.*
- *Neither do the current commercial GM crops have any biologically significant increased levels of endogenous allergens compared to the corresponding conventional crops.*
- *Four allergy tests currently used have no thorough scientific basis.*
- *Recent research on GM crops has shown the misleading nature of these four tests.*

They conclude that the current safety assessments according to Codex guidelines (Textbox 8.1) are based on the currently available knowledge on food allergens and risk, and that they are therefore suitable for assessing the potentially increased risk of allergy from a gene crop compared to the corresponding conventional crop. However, making the four non-scientifically validated allergy tests compulsory may lead to the rejection of safe and useful products, astronomical costs and to a possible cessation in trading without a corresponding reduction in risk. Worse still, the use of unsuitable tests such as unvalidated animal models instead of highly suitable tests could even lead to the introduction of a product that *does* contain a significant risk for a group of consumers with allergies. Textbox 8.2 shows that the

---

TEXTBOX 8.2.
Genetically modified rice fights allergies.

What if the food we ate fought allergies instead of causing them? A new form of GM rice can, researchers announced in 2009 (Domon et al., 2009). The new transgenic rice designed to fight common pollen allergy appears safe in animal studies. In laboratory studies the researchers fed a steamed version of the transgenic rice and a non-transgenic version to a group of monkeys
every day for 26 weeks. At the end of the study period, the test animals did not show any health problems, in an initial demonstration that the allergy-fighting rice may be safe for consumption, the researchers say.

RICE CAN HELP AGAINST ALLERGIES

world can also turn the other way around.

The reality of more than a decade of consuming GM foods has demonstrated that those brought onto the market to date have not been a direct cause of harm to our health and that the safety assessments have therefore, so far, worked as intended. Yet the scientific basis of some tests is still shaky, as suggested above. Alan McHughen, a professor at the University of California and former chairman of the International Society for Biosafety Research, sums up the "fatal shortcomings" of GM foods legislation in the US in a letter published in 2007 in *Nature Biotechnology* (Anonymous, 2007; McHughen, 2007):

1. *Many legal bodies are still under the false impression that the process of transgenesis (modification with recombinant DNA technology) is inherently risky and that all products developed using recombinant DNA technology must therefore categorically be critically investigated. At the same time, they suppose that 'conventional' breeding processes, including the genome-disturbing methods such as radiation mutagenesis, are risk-free and therefore do not require much or any legal investigation.*

2. *Another misconception is that the combination of genes of different species is unnatural and risky. This causes unnecessary anxiety among consumers and is the reason behind the request for exclusive GMO legislation.*

3. *Products of a comparable risk level should undergo a comparable critical examination.*

4. *Because risk is relative, risk assessments should take place in a context and be compared with alternatives. Yet many recombinant DNA risk analysis protocols only concern the assessments of GMOs, i.e. not in a real-life context. It should be noted here that nothing is 'absolutely' safe, because (1) science can never prove that a product will never cause harm, and (2) everything comes with a certain degree of risk.*

5. *An analysis conducted with technical expertise is not enough to give solid scientific backing to the work as a whole; the reason for conducting the analysis at all, should also be scientifically valid.*

6. *In the case of risk assessments most GMO data requirements are excessive, in any case far beyond the point that is necessary to be able to determine relative safety.*

7. *Political, ethical and economic factors play a role in many risk assessments, thereby blurring the scientific focus.*

In fact we can conclude that (1) eating genetically modified food has thus far caused no direct harm to health, (2) from a scientific point of view there is no justification for testing GM foods more thoroughly than other new foods, (3) further scientific justification of risk analyses of food is required in general, and (4) there should be more worldwide uniformity in risk analyses.

Here's a delicate question we can ask ourselves: would I dare eat GM foods and give it to my children? The answer is yes, and moreover, we are undoubtedly doing so already! Complete separation of conventional and transgenic animal feed, just to give an example, is certainly not yet watertight and is a relatively expensive procedure, so most of our animal products come from cattle that have been fed some GM foods. Statements from representatives of the grain trade and the cattle feed, food and biotechnology industries corroborate this. They point to the fact that the growing proportion of GM corn and soya in Argentina, Brazil and the US makes it increasingly difficult to separate the normal, conventionally cultivated corn and soya from GM corn and soya. It is also becoming more and more difficult to keep GM crops permitted by these countries, but not authorised by the EU, from entering the export channels. This results in higher costs for those GM-free products which the European market is demanding. The industry and trade is therefore appealing for a more lenient European authorisation policy. The downside is that a considerable number of consumers in the EU and US still have doubts about the quality of GM foods. Environmental and development organisations therefore continue to mount serious campaigns against the cultivation of these gene crops.

8.4. MORE ANXIETY!

There is further concern from organic farmers, most of whom think that transgenic crops are not organic, even if they are more pest and disease resistant, and thus require less pesticide. They are afraid that their crops will be "contaminated" by cross-pollination with the genes from the transgenic crops. They believe they are entitled to protection against this, in a passive sense by keeping the different fields at an adequate distance from each other and in an active sense by

erecting barriers (co-existence). EU policy-makers in all member states are struggling with the implementation of coherent co-existence regulations. Demont & Devos (2008) appeal for flexible co-existence regulation which explicitly takes account of economic motives.

NB: Organic agriculture does not by definition produce healthier products than conventional and transgenic crops. In a report published in 2007 by Andrew Staehelin and David Christopher of the American Council on Science and Health, it was shown on the basis of yet more scientific publications that GM crops and food from those crops is even safer and more healthy than "organic" food[56].

ORGANIC FOOD ISN'T NECESSARILY SAFER THAN GM FOOD

Another concern among organic farmers is an accelerated increase in the resistance of pests to the few pesticides they can use. Take, for example,

[56] www.acsh.org/printVersion/hfaf_printNews.asp?newsID=962

crops like cotton and corn genetically modified with a gene from the bacterium *Bacillus thuringiensis*. These transgenic plants then make a protein (Bt toxin) that is toxic for insects, thus preventing them from decimating the crops. The fear is that large-scale cultivation will increase the risk of these insects becoming pesticide-resistant. If, for instance, the bollworm should become resistant to the Bt toxin, it would be a disaster for organic cotton growers, because bacterial Bt toxin is the only means they have to eliminate this insect. In America the mandatory solution is the so-called *refuge strategy*. This theory assumes that a resistant mutant of the insect, which undoubtedly occurs once, has no advantage in relation to the non-mutated insect if the transgenic crop is sufficiently alternated with the normal variant (the refuge). It sounds simple, but in practice it isn't. Questions such as "what percentage should be refuge area?", "What is the best spatial division?", and "What should refuge crop rotation look like?" are not easy to answer. So far the refuge strategy is working, but there is still some doubt as to whether that will continue to be the case as more of these transgenic crops are grown.

Then there is the additional concern that antibiotic-resistant genes, used particularly in the first-generation transgenic crops as selection markers, will be transferred to bacteria, causing them to become resistant to the antibiotics in question as well. There is no proof that this can actually happen, but as a precaution alternatives are currently being used, for example, a gene from a jelly fish, whereby a fluorescent protein is formed that can be seen, and thus its presence detected, under UV light.

Finally, a concern about the complete dependence of agriculture on a limited number of multinationals, endangering the livelihood of small third-world farmers in particular. Anti-cartel laws should in theory prevent this. And the last 30 years have shown that Third-World small farmers are able to adopt more efficient technologies and have in fact done so. Nevertheless, gene technology has made agriculture more dependent on a smaller number of large companies and we believe it is right to ask whether this should continue.

8.5. WHO IS TELLING THE WHOLE TRUTH?

It should be obvious by now that we are not members of the lobby that opposes gene technology. As we have already mentioned in Chapter 1, we have been biotechnologists for decades. Yet, in this book, we are endeavouring to provide both sides with as much objective information as possible. In this section we will show that neither party tells the whole truth all of the time, which can be very misleading for the general public. For the sake of balance and not desiring to lay it on too thickly, i.e. with a view to winning over one party or another, we use just a few examples from each side. There are far more detailed examples in the above-mentioned report by Rader.

Proponents like to say that genetic modification is as old as agriculture itself, thereby implying that there is little difference between traditional breeding and breeding with the help of recombinant DNA technology. It is certainly true that unexpected genetic changes occur in the more conventional techniques, which itself should be an argument for the strict regulation of every newly introduced crop. It can certainly be no valid argument for the safety of transgenic crops to exempt them from careful testing.

The big companies, who have thus far developed the most important transgenic crops, like to use Golden Rice as their model. Golden Rice was developed with the aim of tackling vitamin A deficiency in large parts of the world. This shortage causes many children to go blind, die prematurely or contract all kinds of other diseases. The vitamin A content in Golden Rice is dramatically increased by genetic modification. These big companies have, however, done little to develop this transgenic rice, except for allowing their patents on it to be used. This will yield little, however, since the stakeholders, i.e. the poor farmers in the Third World,

have no money to buy the seed. More about Golden Rice in Textbox 8.3.

An issue about which both parties blatantly exaggerate in our view is food shortages. According to its proponents, biotechnology is the solution to the world's hunger problems, while according to its opponents there is enough food, and the problem is simply one of distribution and politics. The truth lies somewhere in-between, particularly in the future.

And now the Pusztai affair. Árpád Pusztai (1931) is a researcher born in Hungary, who has worked for most of his career at the Rowett Research Institute in Aberdeen, Scotland. He is regarded as a leading expert in plant lectins, an area in which he has published prolifically. In 1998 Pusztai claimed in a BBC programme that the results of his research showed that rats fed with a diet of genetically modified potatoes developed deformed

---

TEXTBOX 8.3.
Golden Rice.

Every year more than two million people go blind. In 60% of cases in India, China and sub-Saharan Africa this is the result of a vitamin A deficiency. Education in the area of health care, vitamin supplementation, gardening, food and feeding programmes and the use of genetically modified rice with higher levels of provitamin A (β-carotene) are possible ways of preventing this. The UN Standing Committee on Nutrition stresses the need to integrate these measures in order to tackle food shortages[57].

[57] www.unsystem.org/scn

Genetically modified rice, known as Golden Rice because of its colour, was developed with the aim of increasing levels of provitamin A. Golden Rice is regarded as the first example of a transgenic crop that had a direct benefit for the consumer, but in fact the first was the tomato Flavr Savr (see Textbox 2.2 in Chapter 2). When Golden Rice was introduced to Asia, people were confronted with the same problem that we might expect with transgenic sorghum in Africa (Botha & Viljoen, 2008). Namely, concerns about the environment, patents, efficacy and social acceptance. As far as the environment is concerned there is a fear that the recombinant genes will jump over to traditional and wild rice varieties, which according to ecologists could have far-reaching consequences for Asia, because rice has its origins there. Golden Rice is reportedly free from royalties, but various international patents on it are in the hands of multinationals. These businesses have agreed that poor farmers don't have to pay royalties if they earn less than 10,000 American dollars from Golden Rice, and if the rice is not exported[58]. In practice, however, it would be difficult for poor farmers to prove compliance with either condition, simply because they don't have the means to do so.

There is also some doubt as to the efficacy of Golden Rice in preventing vitamin A deficiency. Provitamin A (β-carotene) first needs to be converted into retinol, the form of vitamin A that is absorbed by the body. The result is that, at the very most, ten percent of the provitamin A is finally available as vitamin A. This means that you would have to eat Golden Rice to the equivalent of 250 g of uncooked rice per day in order to obtain the required quantities of vitamin A. Conversion of β-carotene and uptake of retinol requires the presence of lipids, especially unsaturated fatty acids, that are not soluble in water. Brown rice contains both β-carotene and the required fatty acids in the innermost layers of the husk. However, this is lost when the husk is removed to make white rice (which is the most popular). In Golden Rice β-carotene is also present in the innermost white part, the endosperm, but not the fatty acids required for conversion, so the efficacy is somewhat reduced. In addition, since Golden Rice is yellow in colour,

there is some question as to whether it will be as socially unacceptable as brown rice due to cultural preferences[59]. Stein et al (Stein, Sachdev, & Qaim, 2006) conclude that Golden Rice can help alleviate vitamin A deficiency, but a range of other approaches is also necessary to tackle this complex problem. They also conclude that the uptake of β-carotene from Golden Rice with the highest content still needs to be scientifically verified. In short, it is still very questionable whether the desired objectives can be achieved with Golden Rice. We refer to the previously mentioned Rader website for a more detailed overview of the fierce battle that has already been waged between opponents and supporters on this subject.

An announcement in the 4 February 2008 issue of AgraFood Biotech (Anonymous, 2008b) gives us hope. According to Robert Zeigler, general director of the International Rice Research Institute (IRRI), farmers will probably be able to get hold of Golden Rice by 2011. By the end of January 2008 he reported that the first field trials would take place that year in the Philippines with the aim of releasing the new crop in 2011, 10 years after the first developments. He was speaking after having received a 20 million dollar subsidy for the project from the Bill & Melinda Gates Foundation. IRRI believes this subsidy can help them supply 18 million households, primarily in South Asia and sub-Saharan Africa, with better rice varieties, while it is expected that the yield will rise by 50% in the next decade.

---

[58] www.econexus.info

[59] www.panap.net

organs. The activities of Greenpeace in 1996 on the quayside at Rotterdam when the first ship laden with transgenic soya arrived had already fuelled the fire of the biotechnology debate, but this revelation really set the cat among the pigeons. Pusztai was fired by the institute shortly thereafter. We believe that this affair has been blown up so much in Europe, thanks to the convenient manipulation of the press by extreme opponents (crash courses in how to manipulate the press are ten-a-penny on the net), that many people have developed an irrational fear of biotechnology and Europe is now lagging seriously behind in this area. Matters have been made worse by the ban on all GM crops which was in force for years after the Pusztai affair.

HOW TO MANIPULATE THE PRESS

Up until 1998 Pusztai worked at the Rowett Institute, where he conducted experiments with transgenic potatoes. These potatoes were genetically modified with the lectin gene from snowdrops. Lectins are a specific type of protein that are present in all nuts, seeds and bulbs. Some of these lectins, such as that in the red kidney bean, are poisonous to humans. Lectins are destroyed at higher temperatures, which is why we first need to cook beans before eating them. Others, such as tomato lectins, are evidently harmless when present. Some lectins are toxic for insects and are therefore seen as a potential natural pesticide. The lectins from snowdrops are toxic for insects, but not for people. In his experiment, Pusztai fed these transgenic potatoes to one group of rats and normal potatoes with added lectins to a control group; both types of potato were eaten raw. Both groups of rats developed deformed organs, and there was no statistically significant difference between the two groups; independent statisticians later confirmed this. However, Pusztai claimed that the organs of the rats that had eaten the transgenic potatoes were *more* deformed and published this research together with Ewen in 1999 in *The Lancet*. He stuck to this claim whenever the press was present. Opponents still use this as an argument against transgenic crops, because it was published in such a renowned journal.

In June 1999 the influential British Royal Society published a critical analysis of Pusztai's results, and concluded that they were not significant for the following reasons:

1. *Poor experimental design, possibly exacerbated by lack of 'blind' measurements resulting in unintentionally biased results.*
2. *Uncertainty about the differences in chemical*

composition between strains of non-GM and GM potatoes.

3. Possible dietary differences due to non-systematic dietary enrichment to meet Home Office and other requirements.

4. The small sample numbers used in experiments testing several diets (all of which were non-standard diets for the animals used) and which resulted in multiple comparisons.

5. Application of inappropriate statistical techniques in the analysis of results.

6. Lack of consistency of findings within and between experiments.

However, the following was also concluded:

"Although we have no evidence of harmful effects from genetic modification, this of course does not mean that harmful effects can be categorically ruled out. This issue can be resolved only by the necessary research carried out to a high standard and by full use of the regulatory mechanisms for dealing with safety of food."

Pusztai himself also emphasised that his findings were preliminary and should be seen as a forerunner to further research.

In the article by Miller *et al.* (Miller, Morandini, & Ammann, 2008) there was also a great deal of attention devoted to the Pusztai affair. They wondered how it was possible for such an inaccurate and incomplete study to have got past the peer review system of such a high-standing journal. The editors of *The Lancet* argue that despite the admittedly lax methodology - and disregarding the serious reservations of the referees - they had published the article with the hope of making a constructive contribution to the debate between scientists, media and the public on a heavily politically-charged subject[60]. Miller *et al.* regard it as a scandalous and irresponsible deceit and a travesty of the peer review system of research articles. We have no choice other than to conclude that in this case the Pandora's box was recklessly opened. We would however prefer to confine ourselves to the conclusion by the Royal Society that only high-quality research within legal frameworks is good enough when it comes to food safety. Neither side can argue with that, and hopefully they will work together on this.

8.6. IS THERE A FUTURE FOR TRANSGENIC CROPS?

"Fear of biotechnology is perpetuated by activists spreading propaganda that is based on zero science … it is the misinformed informing the uninformed."

Ken Hobby, President of the US Grains Council.

We are in no doubt that genetically modified crops have a future, even in Europe, and hopefully also in the Third World. In countries such as the US, Canada and Argentina transgenic food crops have been cultivated

[60] news.bbc.co.uk/2/hi/science/nature/472192.stm

on a large scale for years. In fact, these crops have nothing but advantages for the farmer, and in particular for the multinationals who supply the seeds and agricultural chemicals. What we need are examples where there is a clear benefit for consumers and where it is demonstrated that the crops are as safe as all our other food in all aspects. If we continue to develop this technology carefully, then the future will bring food crops that give a bigger yield, are more nutritious, healthier, tastier and better for the environment. Of that we are convinced. A nice example of this sort of development is genetically modified carrots with higher levels of absorbable calcium. Morris *et al.* (J. Morris, Hawthorne, Hotze, Abrams, & Hirschi, 2008) describe food studies in which people and mice are fed with these transgenic carrots. They show that the uptake of calcium rises considerably compared to control groups fed with normal carrots. Calcium deficiency is a major problem in our Western world, causing osteoporosis especially in the elderly. This deficiency can be

---

TEXTBOX 8.4.

Chopping onions without tears.

Using a special technique called gene silencing, researchers from the Crop & Food Research Institute in New Zealand have deactivated a gene in onions. As a result, these genetically modified onions can be chopped without inducing tears. The deactivated gene codes for the enzyme lachrymatory factor synthase (LFS). When traditional onions are cut this enzyme comes into contact with the sulphur compounds that exist in the onions and these are then converted into 'tear inducers'. In the genetically modified onions, LFS is not made and thus neither are the "tear inducers".

The researchers even anticipate that this will also enhance the taste and nutritional values of these onions. The research director Colin Eady says: "What we're hoping is that we'll essentially have a lot of the nice, sweet aromas associated with onions without that associated bitter, pungent, tear-producing factor. This is an exciting project because it's consumer orientated and everyone sees this as a good biotechnology story."

He expects that in about ten years time the first 'no-tear onion' will appear on the market. His colleague Meriel Jones of Liverpool University, who designed the Suprasweet onion by growing it on low-sulphur soil, hails this research: "This is a great development. It shows how genetic engineering can lead to real benefits for both cookery and health. Although conventional growing has identified some sweet, mild onions, this discovery will eventually give farmers new varieties and consumers more choice."

ONIONS WERE USED FOR KEEPING YOUR AIRWAYS OPEN AT NIGHT WHEN HAVING A COLD

GREAT... "NO-TEAR ONIONS"

RRRRR... ZZZZZ...

---

prevented by using genetic modification to raise the levels of absorbable calcium in fruit and vegetables that contain relatively little calcium. Another more remarkable example of such developments stems from New Zealand and is described in Textbox 8.4 (Anonymous, 2008b).

To date no damage to the environment has been observed as a result of the cultivation of transgenic crops. In the case of transgenic-cotton plantations in particular there has been a drastic reduction in the use of pesticides. In short, if it is possible for opposing parties to tread the DNA path together, then there is, in our view, a "golden (rice) future". So it's a pity that high-ranking people like Ken Hobby still make imprudent statements about the other side (see focal point 2 in Chapter 3). This will do little to unite the two parties: quite the contrary. It is all the more regrettable because this man also does and says some very sensible things. The quote comes from a press release from early 2008 in which the *US Grains Council* (USGC) launched a publicly accessible multimedia and interactive CD-ROM about genetically modified crops available on their website[61]. In the same press release, Hobby also says: "The lack of user-friendly and accessible scientific information in one place is one of the primary reasons why modern agricultural biotechnology still provokes concerns among many consumers and among international legislators. The USCG has made this CD-ROM in an attempt to address this." This touched a chord with us. It is precisely for these reasons that we wrote this book.

[61] grains.org/multimedia/index100.html

Together, we hope we are heading for an informed, safe, healthy and sustainable society.

8.7. CONCLUSIONS

Researchers around the world are studying how to improve crops and farming techniques to address worldwide hunger. By breeding staple crops such as wheat, rice, maize, soya, and sorghum to be more pest- and weed-resistant, more nutrient-rich and high-yielding, and more digestible, they hope to offer more nutrition per hectare of farmed land. "Grain sorghum is a very important crop in Africa", Kent Bradford, director of the Seed Biotechnology Center at the University of California, Davis, said in an interview with Clara Moskowitz of LiveScience (Moskowitz, 2008). "Unfortunately, its protein content is relatively indigestible – the nutrient is inefficiently metabolized. There is work in trying to modify sorghum so the protein is more digestible. That would be a huge bonus." Some scientists think the real key to ending world hunger lies in genetically modifying crops to provide boons that nature cannot match. But many people question the wisdom of dabbling in complicated natural processes that we don't fully understand. "I think using genetically-engineered crops would not only not solve the situation, but it would continue to put the food supply at risk," said Ryan Zinn, campaign coordinator for the Organic Consumers Association, a non-profit organization, in the same article in LiveScience. "When you're messing with the crop's genome, you run the risk of opening Pandora's box." Conversely

Bradford told LiveScience: 'Nobody can point to a single thing to say there's been unintended health consequences. While it's always possible, it's also possible that breeding crops could have unintended health consequences. It's a matter of balancing risks and benefits. The risks are exceedingly small, but the benefits tangible.' We fully agree with Bradford and repeat focal point 7 of Chapter 3 stating that only integrated approaches can help Third-World countries, not just genetic engineering.

8.8. SOURCES

Anonymous. (2007, 23 July). McHughen lists GM regulation's "fatal flaws". *AgraFood Biotech,* p. 9.

Anonymous. (2008a, 17 March). Ninety-day feeding studies satisfactory, says EFSA. *AgraFood Biotech,* p. 4.

Anonymous. (2008b, 4 February). No-tear onions. *AgraFood Biotech,* p. 21.

Botha, G. M., & Viljoen, C. D. (2008). Can GM sorghum impact Africa? *Trends in Biotechnology, 26*(2), 64-69.

Demont, M., & Devos, Y. (2008). Regulating coexistence of GM and non-GM crops without jeopardizing economic incentives. *Trends in Biotechnology, 26*(7), 353-358.

Domingo, J. L. (2007). Toxicity studies of genetically modified plants: A review of the published literature. *Critical Reviews in Food Science and Nutrition, 47*(8), 721-733.

Domon, E., Takagi, H., Hirose, S., Sugita, K., Kasahara, S., Ebinuma, H., & Takaiwa, F. (2009). 26-Week Oral Safety Study in Macaques for Transgenic Rice Containing Major Human T-Cell Epitope Peptides from Japanese Cedar Pollen Allergens. *Journal of Agricultural and Food Chemistry, 57*(12), 5633-5638.

EFSA. (2008). Safety and nutritional assessment of GM plants and derived food and feed: The role of animal feeding trials. *Food and Chemical Toxicology, 46*(Supplement 1), S2-S70.

Goodman, R. E., Vieths, S., Sampson, H. A., Hill, D., Ebisawa, M., Taylor, S. L., & van Ree, R. (2008). Allergenicity assessment of genetically modified crops - what makes sense? *Nature Biotechnology, 26*(1), 73-81.

Kuiper, H. A., Kleter, G. A., Noteborn, H., & Kok, E. J. (2002). Substantial equivalence - an appropriate paradigm for the safety assessment of genetically modified foods? *Toxicology, 181,* 427-431.

McHughen, A. (2007). Fatal flaws in agbiotech regulatory policies. *Nature Biotechnology, 25*(7), 725-727.

Miller, H. I., Morandini, P., & Ammann, K. (2008). Is biotechnology a victim of anti-science bias in scientific journals? *Trends in Biotechnology, 26*(3), 122-125.

Morris, J., Hawthorne, K. M., Hotze, T., Abrams, S. A., & Hirschi, K. D. (2008). Nutritional impact of elevated calcium transport activity in carrots. *Proceedings of the National Academy of Sciences of the United States of America, 105*(5), 1431-1435.

Morris, S. H. (2007). EU biotech crop regulations and environmental risk: a case of the emperor's new clothes? *Trends in Biotechnology, 25*(1), 2-6.

Moskowitz, C. (2008, 23 April). Radical Science Aims to Solve Food Crisis. *LiveScience*

O'Riordan, T., & Cameron, J. (1994). *Interpreting the precautionary principle.* Earthscan/James & James.

Stein, A. J., Sachdev, H. P. S., & Qaim, M. (2006). Potential impact and cost-effectiveness of Golden Rice. *Nature Biotechnology, 24*(10), 1200-1201.

van Haver, E., Alink, G., Barlow, S., Cockburn, A., Flachowsky, G., Knudsen, I., Kuiper, H., Massin, D. P., Pascal, G., Peijnenburg, A., Phipps, R., Poting, A., Poulsen, M., Seinen, W., Spielmann, H., van Loveren, H., Wal, J. M., & Williams, A. (2008). Safety and nutritional assessment of GM plants and derived food and feed: The role of animal feeding trials. *Food and Chemical Toxicology, 46,* S2-S70.

part three

Health has limits

"Recently there has been much criticism of health care, but since 1950 infant mortality has declined by a factor of five and the average life expectation has increased from 71 to almost 80 years."

Hans Galjaard

Emeritus Prof. Hans Galjaard is the father of prenatal diagnostics and clinical genetics in the Netherlands. Not only is he a proficient physician, he is also the author of the best-selling *Alle mensen zijn ongelijk* (All men are different) and of the book published in 2008, *Gezondheid kent geen grenzen* (Health has no limits). He is also the man behind the statement: "It's fascinating how many new insights have been gained into the evolution of plants, animals and humans, thanks to the genetic revolution." The genetic revolution is also expected to cause a fascinating turnaround in health care. Prior to the genetic revolution there were three other developments that transformed health care. These were: better hygiene as a result of the introduction of sanitation systems and sewers; anaesthesia, which enabled doctors to treat sick patients under sedation; and finally the introduction of vaccines and antibiotics, which prevented and treated many infectious diseases caused by viruses and bacteria. Antibiotics and the genetic revolution are the focus of part III of this book. The possibilities in health care seem particularly boundless as regards the genetic revolution. It is also clear that the road to a panacea, to a drug to cure all ills, has still not reached its end. For the time being, therefore, health does have limits!

HYGIEIA   PANACEA   ASCLEPIUS

The Greek God of medicine and healing Asclepius with his daughters Hygieia (goddess of health) and Panacea (goddess of healing), all three accompanied by the snake, the symbol of health.

There is a tide in the affairs of men. Which, taken at the flood, leads on to fortune.

This quote from the play Julius Caesar by Shakespeare sums up better than any other the history of penicillin, a substance excreted by the mould *Penicillium notatum*. Penicillin was discovered in 1928 by the British scientist Alexander Fleming. A series of compounds, belonging to one of the most important categories of antibiotics, was derived from this substance. The enormous potential of this antibiotic, which first became obvious during the Second World War, has been realised in all kinds of ways. Innumerable human lives have been saved by using this category of antibiotics in cases of bacterial infection. But that's not all. The development of the penicillin production process was a great stimulus to the progress of modern biotechnology in general and of large-scale biotechnological processes in particular (Demain & Elander, 1999; Demain & Sanchez, 2009; Tramper *et al.*, 2001). Bacteria are, however, tough little rascals and can quickly build up resistance. So there arose a sort of eternal "arms race" between bacteria and antibiotics. Modern biotechnology enables new weapons to be developed for preventative and curative purposes, and these are also deployed against a whole range of other diseases. A few notable developments in the field of antibiotics are the subject of this chapter.

9.1. ANTIBIOTICS: LIFE-SAVING BIOTECHNOLOGY

Since its discovery in 1928 many a book has been written about penicillin. This substance, which has had a major impact on the course of history, is referred to as "yellow magic" by some writers, even though it isn't actually yellow in its pure form. The story behind the discovery of penicillin by Sir Alexander Fleming is well worth repeating. Fleming was a scientist studying the growth of bacteria. One day one of his bacteria cultures, a *Staphylococcus*, which he was growing in Petri dishes, was contaminated by a mould, later identified as *Penicillium notatum*. A few days later, before cleaning the dish, he took another look and discovered that all the bacteria around the mould had disappeared (Figure 9.1). It is to Fleming's great credit that he realised the significance of this observation, namely that a substance (named penicillin by him, because it was produced by a *Penicillium* mould) hindered the growth of bacteria. It was this discovery that finally earned him a Nobel Prize for Medicine in 1945. However, he made little contribution to developing it into a medicine. This was done by Howard Florey and Ernst Chain who shared the Nobel Prize with Fleming (Landsberg, 1949).

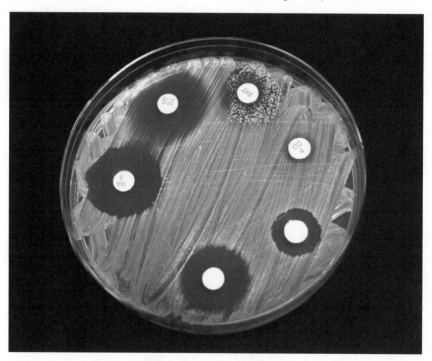

Figure 9.1. Petri dish with fungus (white circles) and bacteria cultures (smears). (Source: Fotolia).

It wasn't until a decade later that Alexander Fleming's discovery that bacteria cultures disappeared in Petri dishes accidentally contaminated by moulds resulted in a pharmacological revolution: the synthesis of pure penicillin by Florey and Chain from the moulds that Fleming described. The story of how Florey isolated the first pure penicillin at the University of Oxford is still one of the most gruesome stories in the history of pharmacy. The moulds needed for the manufacturing seemed to thrive best in dirty hospital bedpans. At first it seemed impossible to get the antibiotic in pure form from this 'raw material'. Florey solved the problem by administering the thus-acquired moulds orally to the local police force, which was thereby collectively transformed into a human bioreactor before the term even existed. Pure penicillin was obtained from the urine of the fearless officers. When introducing his discovery in the United States, Florey truthfully declared to the amazement of his American colleagues: "It's a remarkable substance, gentlemen: grown in bedpans and purified by the Oxonian police force" (Anonymous, 1995).

Penicillin was first used on a large scale in the American field hospitals in the Second World War, and led to an astonishing reduction in mortality from infectious diseases. American and Canadian soldiers introduced it to the Netherlands. Within a few years death from bacterial infection fell to unprecedented

THE FIRST HUMAN BIOREACTOR...
WAS A POLICE FORCE IN OXFORD!

AHHH, IT FEELS GREAT TO HELP!

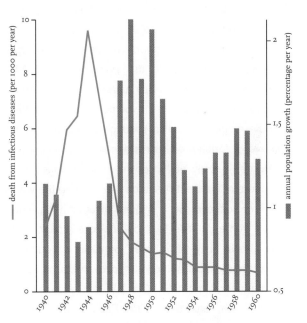

Figure 9.2. Penicillin, infectious mortality and the post-war population growth (adapted from Anonymous, 1995).

levels in the Netherlands too (Figure 9.2). The main problem initially was how to meet the surge in demand. First the moulds were grown in large quantities in 'milk churns'. Later, ever bigger bioreactors were designed. This technology facilitated the development of many other large-scale biotechnological processes.

In the Second World War the Dutch Yeast and Spirit Factory in Delft was also working in deep secrecy on the development of a production process for penicillin. Five years after the war the now Royal Dutch Yeast and Spirit Factory became one of the biggest penicillin producers in the world. The British science historian Marlene Burns wrote a thesis (Burns, 2005) on this development, which reads like a pure-bred thriller. She obtained her doctorate on the strength of it in 2005 at the University of Sheffield (*The development of penicillin in the Netherlands 1940-1950: the pivotal role of NV Nederlandsche Gist- en Spiritusfabriek, Delft)*. After the war the Delft factory evolved into one of the world's biggest penicillin producers because of a mixture of fortuitous circumstances. But, according to Burns, it really boiled down to what united the members of the Delft research team during the war: *a will to succeed*. So it is quite disappointing, she writes with a suitable feel for *understatement*, that now almost sixty years later, penicillin production in Delft has almost entirely stopped. In March 2005 the management of DSM Gist BV (successor to the Royal Dutch Yeast and Spirits Factory) decided to move most of its production to China and India.

In the beginning penicillin was a first-rate remedy against bacterial infections and was often life-saving. However, all kinds of bacterial strains soon became resistant to it. This was mainly, but not exclusively, the result of the exchange of plasmids with resistant genes (Textbox 2.1 in Chapter 2). Mobile DNA (so-called *jumping genes*), bacterial viruses and the uptake of material from dead bacteria are also means used by bacteria to pass on resistance to each other. In the same year that penicillin was introduced on an industrial scale, resistant bacterial strains were appearing, in particular among staphylococci, which are normal skin bacteria that can cause horrible wound infections. Eight years later penicillin was only effective against 15% of *Staphylococcus aureus* infections. *Staphylococcus aureus* is the notorious bacterium which is now virtually resistant to antibiotics and has led to whole hospital departments being shut down for varying periods of time; it is referred to as MRSA, short for methicillin-resistant *Staphylococcus aureus* (methicillin is a second generation antibiotic derived from penicillin) or, these days, multi-resistant *Staphylococcus aureus*, because this strain has become resistant to virtually all antibiotics. Now not only are the staphylococci multi-resistant, but so are, for example, pathogens like *Klebsiella*, *Serratia* and *Acinetobacter* strains, causing among others infections of respiratory and urinary tracts.

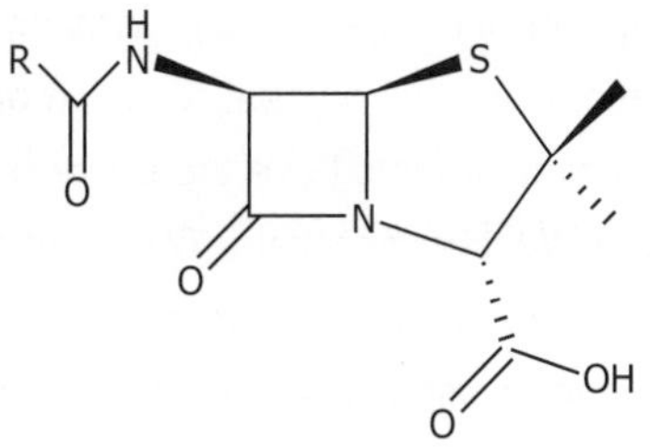

Figure 9.3. Basic structure of penicillin antibiotics.

In 1959 second generation antibiotics came onto the market, the so-called semi-synthetic antibiotics. Penicillin consist of a nucleus and a side chain (Figure 9.3). With the help of chemical synthesis the side chain R of penicillin can be split off and a selection of other side chains can be connected to the nucleus in its place. This was the start of a whole new generation of antibiotics to which bacteria were not then resistant. Two years after their introduction there were again a few *Staphylococci aureus* resistant to these new antibiotics. It seemed to be only a question of time before staphylococci would develop resistance against practically all new antibiotics. Only the so-called glycopeptide antibiotics - most importantly among them vancomycin and teicoplanin - remained free from resistance. This type of antibiotic has a different mechanism of action from the penicillins. Only one gene is involved in resistance in penicillins, whereas there are at least five involved for resistance against the glycopeptide antibiotics.

In the 1980s people were convinced that the pharmaceutical industry was winning the battle against resistance. However, this conclusion seemed a little premature. In 1997 vancomycin and teicoplanin also finally failed as the antibiotics of last resort when resistant bacterial strains were discovered. If the spread of these resistant strains is not stopped, the floodgates will open all over the world. The pharmaceutical industry, but particularly smaller biotechnology companies, are therefore urgently looking for alternatives, especially since the development of a new medicine is a lengthy process (Textbox 9.1).

TEXTBOX 9.1.

Phases in drug development.

A drug undergoes approximately twelve to fifteen years of research before it appears on the market. This time period includes the following phases:

Research phase (3-5 years)
New substances are made and tested on their efficacy.

Preclinical trial (2-3 years)
First in vitro (including tissue culture), then on animals.

Clinical trial (5-7 years):
   Phase I Healthy volunteers are administered varying doses.
   Phase II A small group of patients receive the drug. The therapeutic effect is compared with existing medicines.
   Phase III As in phase II, only on a bigger scale and for a longer period.
   Phase IV Once the drug has appeared on the market, additional clinical research takes place among patients using the drug.

THE BACTERIA FIGHT BACK

In 2007 a very good book by Annet Mooij was published on this matter (Mooij, 2007). It is called *De onzichtbare vijand – Over de strijd tegen infectieziekten* (The invisible enemy - On the fight against infectious diseases). According to Mooij, if there is any lesson to be learned from the last decade it is this: that every advance in the fight against infectious diseases is conditional. The major plagues from the past appeared to be under control, but diseases like malaria, cholera and tuberculosis are now no longer on the way out but on the way back. In the 21 February 2008 issue of *Nature*, the article "Global trends in emerging infectious diseases" by Jones *et al.* (2008) states that most of the (returning) upcoming infectious diseases are caused by bacteria, which is a clear reflection of the great numbers of recorded bacteria resistant to antibiotics. Mooij also writes: "In contrast to the belief of about forty years ago, the book of infectious diseases will never be closed. The widespread use of antibiotics, unnecessary treatments and half-finished treatments means that antibiotic resistance is becoming an increasingly big problem. These practices create an environment for microorganisms in which resistant variants have the upper hand. This is the case in Western hospitals. The most notorious hospital bacterium (MRSA) has been on the rise since the 1980s, and is especially prevalent abroad. The situation in Dutch hospitals is not so bad because of the strict antibiotic policy adopted here. Modern pig sties are also a breeding ground for resistant bacteria, because pigs are administered antibiotics in massive doses. They were previously given as growth promoters and to prevent infections, but in 2006 this practice was banned and antibiotics were no longer allowed to be added as standard practice to the feed. Nevertheless, antibiotic use in pig farming is still abundant, with the result that about 40% of pigs are infected with MRSA. A quarter of pig farmers is also infected. Now there are staphylococci in circulation that are resistant to all available antibiotics. If these infections can no longer be treated, an old and serious problem will again rear its ugly head: wound infections. An ominous prospect."

Mooij ends her book with the following paragraph:

"In the fight against infectious diseases the human biological machinery is clearly missing the target. Whether we are able to make it in the long term depends on the resources we have at our disposal: new medicines and vaccines, knowledge and ingenuity. The future of humanity and microbes likely will unfold as episodes of a suspense thriller that could be titled *'Our Wits Versus Their Genes',* wrote the microbiologist and Nobel Prize winner Joshua

Lederberg. It will be a tragedy in many acts, an infinite play, but thrilling it will remain."

MODERN PIG STIES ARE A BREEDING GROUND FOR RESISTANT BACTERIA

9.3. THE PROSPECTS

A gloomy outlook, then, as far as infectious bacterial diseases are concerned. Let us therefore summarise the pertinent trendsetting review articles from the first decade in this new millennium and see if these present a more positive picture.

Christopher Walsh put the question in *Nature Reviews Microbiology* of October 2003 (Walsh, 2003): "Where will new antibiotics come from?" He maintains that there was an innovation gap of almost 40 years after 1962, before the arrival of two new categories of antibiotics with different antibacterial mechanisms. But he doubts whether this has closed the gap. As a

second indicator that the clinical pipeline for antibiotics is virtually empty, he points out that few major pharmaceutical companies are active in the arena of antibacterial infectious disease. The departure and/or partial retreat of many pharmaceutical companies from this therapeutic area around 1990, reflects not only a mix of economic and market projections, but also a partial or complete failure of research programmes that used existing models as a basis for finding new *leads* that were robust enough to become clinical candidates. In short, according to Walsh, there are not many new (categories of) antibiotics in the pipeline.

The article "Antibiotics: where did we go wrong?" in *Drug Discovery Today* of January 2005 by Overbye and Barrett (2005), who both work for a major pharmaceutical company, takes an identical line. The authors present an overview of antibacterial medicines that are in the clinical trial phase (Textbox 9.1), e.g. for major pharmaceutical companies three in Phase I, one in Phase II and four in Phase III, and for smaller biotech companies two in the preclinical phase, four in Phase I, three in Phase II, one in Phase II-III, and two in Phase III. They conclude that there has been a shift in antibacterial R&D efforts from the major pharmaceutical companies to a whole contingent of biotech companies. This small business approach has led to an explosion in creativity as regards strategies, choice of targets, use of genomics, and development paradigms. The trends they see include (1) a combination of acquisition of niche products that are not being developed by major pharmaceutical companies, (2) the use of scientific discoveries that have not been applied successfully in the search for medicines by major pharmaceutical companies, and (3) an increasing improvement in an existing category of antibiotics. To their surprise none of the major pharmaceutical companies has successfully developed these new target approaches in order to identify potential medicines.

In their analysis of where 'they' went wrong, there are about eight factors, but they stress that finger-pointing is not the solution. The rapidly rising resistance requires the industry to join forces and undergo far-reaching paradigm shifts in relation to how antibiotics can be developed and brought onto the market. In their view a possible way forward is to admit that "we" have fallen short of the target (by a process of gleaning the lessons learned from each other from previous experiences) and that we must continue to look for a joint, universal solution to convince the pharmaceutical industry to invest again in antibacterial research and development. As to the question "Where did we go wrong?", there is no one answer, but there must be one common solution.

Six months later, in 2005, Barrett (Barrett, 2005) alone posed the following question in *Current Opinion in Microbiology*: "Can biotech deliver new antibiotics?" He reported that after the publication in 1995 of the first two complete genome sequences of two pathogenic bacteria (*Haemophilus influenzae* and *Mycoplasma genitalium)* many academic and industrial laboratories threw themselves into pathogenic bacterial genomics with the aim of developing new anti-bacterial methods and/or antibiotics. This has delivered many spectacular scientific results, but as yet no new potential agents to fight bacterial infections. None of the biotech companies who got onto the band wagon has survived on the basis of its own internal R&D programme. Those who did survive have various common characteristics, e.g. a solid and well-defined work plan, excellent scientific leadership, a unique platform for a known product, heavily financed starting capital and little income. They also have or had the virtually unattainable objective of bringing an internal, potential antibiotic onto the market. The big problem with this is the enormous amount of capital needed for the clinical trials.

As we've already seen, the biotech companies have a dozen antibiotics sitting in the pipeline, and according to Barrett these are to be launched over the next six to seven years. A notable example is ceftobiprole, which is currently the leading compound in a new generation of cephalosporins (Figure 9.6). In March 2008 this compound received the approval letter from the FDA for the treatment of complicated skin infections (Cornaglia & Rossolini, 2009). Barrett expects that really new genomics-based developments will not be implemented until around 2015. However, the average failure rate of medicines in this stage of development means that no more than two to four of these candidate antibiotics will reach the market, and none of them are expected to be a great success. But in Barrett's view it does show that the biotech companies could deliver such products, in isolation from the R&D paradigm, and it is this potential success that will propel the continued investment. Basically, modern biotechnology can deliver antibiotics, but it must be done in collaboration with the major pharmaceutical companies so that the costly clinical trials can be funded.

In October 2003 Walsh asked where the new antibiotics would come from (Walsh, 2003). In December 2006 he and two colleagues (Clardy, Fischbach, & Walsh, 2006) answered this question in *Nature Biotechnology*: "New antibiotics from bacterial natural products." The general gist of their article is as follows. The demand for new antibiotics has largely been met in the last five decades by semi-synthetic customisation of a few natural molecular *template* structures, such as penicillin, which were discovered in the mid-20th century. More recently, however, technological advances, e.g. the introduction of high-throughput screening techniques, have seen a reincarnation of the search for antibiotics from bacterial natural products. The main focus of the search is for new antibiotics from old (e.g. streptomycetes) and new sources (e.g. actinomycetes, cyanobacteria and as yet uncultivated bacteria). This has resulted in various newly discovered antibiotics with unique template structures and/or new mechanisms, with the potential to form the basis of new categories of antibiotics. In their list of these antibiotics which are already in the clinical trial phase, we see practically the same substances that we saw in Barrett's table. In many cases, though, a major pharmaceutical company has come on board. An example of this trend is described in Textbox 9.2.

Yet another interesting review article, called "Will We Still Have Antibiotics Tomorrow?", was published in 2007 (Dronda & Justribo, 2007). What's interesting here is that it was written by the medical specialists Salvador Bello Drond and Manuel Vilá Justribó, who work in the Spanish university hospitals of Zaragoza and Lleida, respectively. Even more interesting is that Spain is one of the countries that is a real breeding ground for resistant bacteria because of the excessive and practically unlimited use of antibiotics.

These two medics are well aware of this and wrote the article as a warning. They claim that because of the fast growth in the number of multi-resistant bacteria, fears are growing among doctors and that this fear is gradually starting to filter through to society as a whole. In their view, the problem is made worse by the fact that very few people are aware that there is little hope at this moment of many new antibiotics coming on the market in the short to medium term. The authors advise that careful use of the available antibiotics, based on a detailed knowledge of their mechanism and the use of new forms of administration, such as *inhalers*, may solve the problem in part. A very sound piece of advice, particularly, but certainly not exclusively, for a country like Spain. The article "How antibiotics can make us sick: the less obvious adverse effects of antimicrobial chemotherapy" from 2004 by Stephanie Dancer in *The Lancet Infectious Diseases* also carried the same message (Dancer, 2004).

Another notable review article on antibiotics we found was called "Combination drugs, an emerging option for antibacterial therapy." It was written by Cottarel and Wierzbowski (Cottarel & Wierzbowski, 2007) from the Center for Advanced Biotechnology, Boston University, and was published in December 2007. The authors

TEXTBOX 9.2.

Vicissitudes of a typical anti-infectives biotech company.

Lepetit Research Centre was until 1995 a medium-sized (100-150 employees) research laboratory, belonging to Marion Merrell Dow, devoted to the discovery and development of novel anti-infectives. It was established in Gerenzano near Milan, Italy. Lepetit discovered rifamycin followed by teicoplanin, ramoplanin, lantibiotic actagardine, A40926 and dalbavancin, all important antibiotics. In 1995 Lepetit was bought by Hoechst and became Hoechst Marion Roussel, representing the pharmaceutical branch of Hoechst. As result of this operation Biosearch Italia arose as a spin-off in 1997. Biosearch Italia presented itself as a small Italian biopharmaceutical company focusing on new antibiotics for the treatment of infections caused by multi-resistant pathogens. They focused on glycopeptides, the class of antibiotics to which vancomycin, teicoplanin, A40926 and dalbavancin belong, and on other inhibitors of bacterial

and fungal cell walls. At that time the company worked together with Wageningen University, the Netherlands, on the cleavage of the side chain of A40926 (Figure 9.4). The product can be used as template to prepare new glycopeptide derivatives (Jovetic, de Bresser, Tramper, & Marinelli, 2003). In 2000 Biosearch Italia became the first small biotech company in Italy to go public and appear on the Nuovo Mercato stock exchange. Then in 2003 it merged with the American biopharmaceutical company Versicor Inc. into Vicuron Pharmaceuticals (listed on both the NASDAQ and Nuovo Mercato stock exchange). At that time the company had three molecules in the clinical pipeline (Phase II and III), i.e. dalbavancin, anidulafungin and ramoplanin. In 2005 the company was bought for a stunning $ 1.9 billion by Pfizer, who brought anidulafungin onto the market as a novel antifungal for systemic infections. By late 2006 Pfizer implemented a global R&D restructuring and closed the research centre in Gerenzano. However, molecules discovered by Lepetit scientist continue their story. Ramoplanin, since December 2009 acquired by Nanotherapeutics,

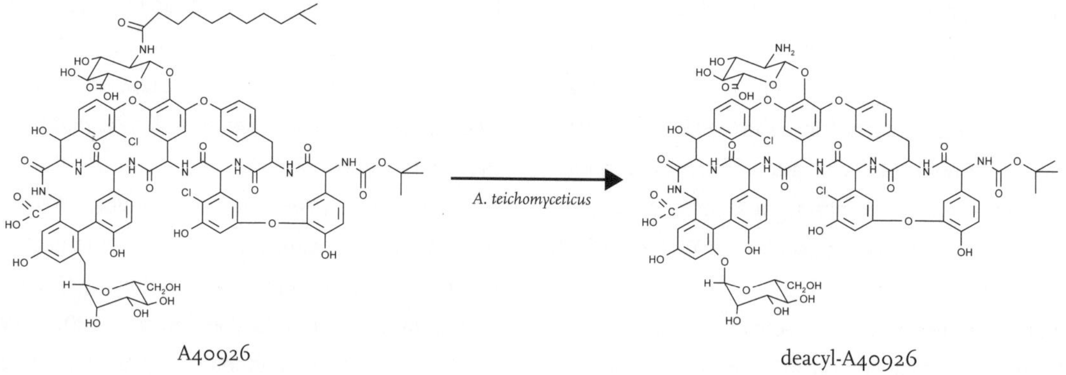

Figure 9.4. Bioconversion of A40926 into deacyl-A40926.

Inc., based in Florida, is now entering Phase III trials as an oral antibiotic for the treatment of Clostridium difficile-*associated disease (Shah et al., 2010). Dalbavancin (Malabarba & Goldstein, 2005), a second generation* glycopeptide, *has been recently acquired by a newly formed US-based biopharmaceutical company Durata Therapeutics, Inc., and is proceeding in its late stage clinical development.*

conclude like everyone else that there is an urgent need for new effective therapies for the treatment of infectious bacterial diseases, whereby the increase in resistance is minimised. In their view, combinations of different categories of antibiotics or the addition of *adjuvants* (pharmacological agents that reinforce the antibacterial action) are a promising alternative therapeutic approach whose efficacy has already been proven in, for example, tuberculosis. Starting with the existing categories of antibiotics (Table 9.1), it is not only possible to increase the activity of well-known and effective antibiotics with combinations, but also to support the development of substances which have already proved to be very effective antibiotics but which are too toxic for patients with a bacterial

Table 9.1. Most important classes of antibiotics and their action.

Class	Action
Fluoroquinolones	Inhibition of DNA synthesis
Aminoglycosides, tetracyclines, ketolides, macrolides, chloramphenicol, lincosamides	Inhibition of protein synthesis
Rifampicin	Inhibition of RNA synthesis
Trimethoprim, sulfonamide	Inhibition of folic acid synthesis
Penicillins, cephalosporins, carbapenems, daptomycin	Inhibition of cell wall synthesis
Colistin, polymyxin	Damage to cell wall integrity

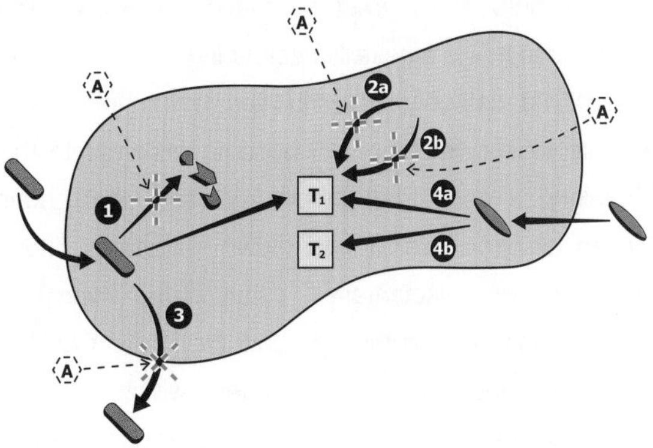

infection. Another advantage of this approach is that it may result in shorter treatment periods and/or lower doses, which may reduce the speed with which pathogenic bacteria become resistant. There are now a few small (start-up) biotech companies working on the development and marketing of such antibacterial combinations.

The authors divide the combination therapies into four categories on the basis of the mechanism of action with which the components potentiate the activity (of each other) (Figure 9.5):

Figure 9.5. Mechanisms of combination therapy: (1) Adjuvant (A) inhibits the degradation or modification of the drug; (2) adjuvant inhibits the cell repair (a) or intrinsic resistance pathway (b); (3) adjuvant inhibits the efflux pumps; (4) combination of two antibiotics with (a) or without (b) similar target T, reproduced with permission (Cottarel & Wierzbowski, 2007).

1. *The breakdown or modification of the actual antibiotic is prevented by a second compound, the adjuvant.*
2. *The accumulation or retention of the actual antibiotic in the cell is facilitated by an adjuvant which stops it being pumped out.*
3. *The intrinsic repair and tolerance mechanism of the bacterial cells against the antibiotic is inhibited by an adjuvant.*
4. *A second compound is itself also an antibiotic with the same or a different mechanism of action from the primary antibiotic.*

Some of these combinations have already been used successfully to fight difficult infections. The most well-known example is Augmentin® which was brought onto the market by GlaxoSmithKline. The antibiotic in this is amoxicillin belonging to the β-lactam (penicillins) category of antibiotics which inhibit cell wall synthesis. Bacteria can become resistant to this by taking a gene from other organisms that codes for an enzyme that breaks down β-lactam. These enzymes, the β-lactamases, catalyse the hydrolysis of penicillins causing them to become ineffective. If a pathogenic bacterium takes up a gene which codes for a β-lactamase that can break down amoxicillin, it has then become resistant to it. Clavulanate, a compound that inhibits these β-lactamases, is present beside amoxicillin in Augmentin®. As a result amoxicillin can still have an antibacterial effect on these resistant bacteria. All in all, not a very rosy outlook, but luckily this Pandora's box is not completely without hope!

9.4. 'GREEN' PRODUCTION

Although it seems from the above that semi-synthetic antibiotics have a limited life, they are still produced and prescribed on a large scale. The "big" antibiotics such as ampicillin, amoxicillin and cefalexin are expected to be in use for a further 20 years. The first two are penicillins, the third belongs to the cephalosporins, which are derivatives of penicillin (Figure 9.6). DSM-Gist is the global market leader in this area. The biggest competition comes from Spain, Italy, India and China. The processes used in these countries start with the fermentation product penicillin G obtained from traditional moulds. This is used to make the semi-manufactures 6-APA and 7-ADCA from which various semi-synthetic antibiotics are then manufactured, including the above-mentioned ones.

In the Dutch DSM production process of cefalexin, which began in March 2001 in Delft (the former Gist-brocades), a genetically modified *Penicillium* strain is used, which dramatically reduces the number of processing steps and thus the costs. The new process also uses 35% less energy and the use of organic solvents has been reduced to virtually zero. By using genetically modified moulds and by replacing chemical synthesis with biocatalysis (i.e. the use of enzymes as biocatalysts) in the subsequent steps too, the production of antibiotics has become much more environmentally friendly; hence the name 'green' chemistry. As mentioned in Section 9.1, all antibiotic production processes at DSM have been transferred to India and China. Only the production process for cefalexin with the genetically modified mould is still carried out in the factory in Delft. DSM is not only the biggest producer of these antibiotics, it is also leading the way in terms of knowledge (generation). A large-scale study in the form of an intensive collaboration between DSM and six Dutch academic research groups over a five-year period (the Chemferm project) ended in 2001 with the publication of a book edited by Alle Bruggink. The book contains a summary of the results of this teamwork and

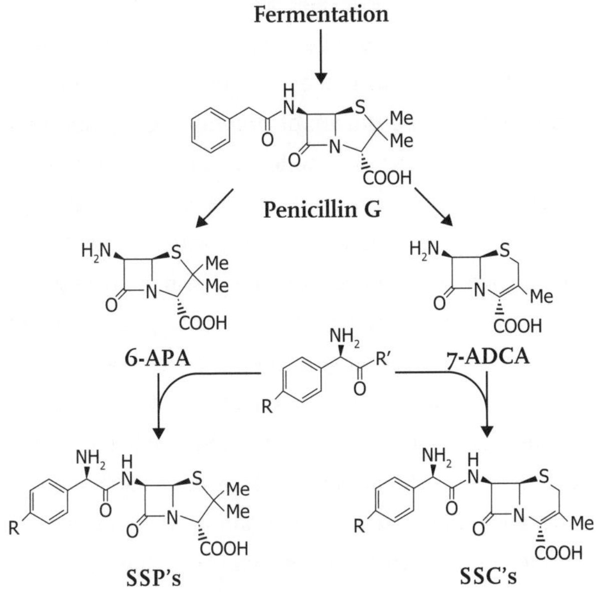

Figure 9.6. General production chart of penicillin-derived antibiotics (semi-synthetic penicillins, SSP's, e.g. ampicillin (R = H) and amoxicillin (R = OH)) and cephalosporin-derived antibiotics (SSC's, e.g. cefalexin (R = H) and cefadroxil (R = OH)). Intermediates: 6-aminopenicillanic acid (6-APA) and 7-aminodesacetoxy-cephalosporanic acid (7-ADCA). Reproduced with permission from Tramper et al. (2001).

consists of more than 100 scientific publications and various patents; Figure 9.6 comes from this book. DSM also conducts a lot of molecular biological and genetic research on moulds and in 2005 they unravelled the gene card, the genome, of *Penicillium chrysogenum*. This is not the strain that Fleming used for his discovery. It is the modern workhorse for penicillin production, and has since been 'bred' so that it produces a thousand times more penicillin than its natural predecessor which was plucked from a melon in 1953. Until recently DSM were the exclusive owners of the genome sequence, but now, 80 years after the discovery of penicillin, they have published it (van den Berg *et al.*, 2008), probably in order to pass the post before other researchers. Either way, it opens the way to yet more innovative processes and antibiotics based on modern biotechnology.

9.5. A NEVER-ENDING STORY

It seems highly likely that the battle between bacteria and antibiotics will become a never-ending story. This seems to hold true today more than ever. New medicines will continue to be followed by new resistances. Global public health then becomes a matter of which is faster, the bacteria or the pharmaceutical industry. Given the time required to build up resistance and to develop a medicine, this looks set to be and is in fact already an exciting race. As we know, a few biotech companies have already started on new experimental methods to tackle resistant bacteria once and for all. Recently we have published a review on this topic (Jovetic *et al.*, 2010). Many pharmaceutical companies have also taken the plunge, literally and figuratively, looking for new types of antibiotic and other biological activities. According to Williams (Williams, 2009) marine bacteria will equip us in the coming century, if properly developed and used, with weapons for our eternal battle against multi-resistant bacteria.

The sea is our richest source of biodiversity and is, so far, practically unexploited. Sponges in particular demonstrate a wealth of biological activity that is promising for medical use. 'Overfishing', however, threatens these very vulnerable ecosystems. It is therefore vital to develop technologies to prevent this occurring. That is why a number of researchers from our own research group are working, for example, on bioreactors, in which they plan to cultivate huge quantities of sponges from very small quantities under tightly controlled conditions; this being the first criterion for developing a similar pharmaceutical process (Sipkema *et al.*, 2005). It is sometimes also worth first taking a look back in history and revisiting old knowledge in the light of what we know now. This is dealt with in the following section of this chapter.

9.6. TAKING ANOTHER LOOK AT PHAGES

Bacteriophages or "bacterial viruses", usually called phages, are natural specialists in killing bacteria. To do this they produce a whole range of antibacterial proteins. These phage proteins may be a source of inspiration in the search for new antibiotics, according to a study by Canadian scientists (Projan, 2004) A phage protein exercises its antibacterial mechanism by binding to a specific location on a bacterium protein which is essential for the survival of this bacterium. The researchers use the interaction between these two as a basis for a method to quickly trace new antibiotics. The researchers speculated that a chemical substance that can interrupt the interaction between an antibacterial phage protein and an essential bacterium protein may well have the same antibacterial effect. After a screening of 125,000 small molecules, a targeted selection of a phage protein and *Staphylococcus aureus* protein provided eleven candidate antibiotics; these indeed seemed to be able to inhibit the growth of *Staphylococcus aureus*. Small molecules are better antibiotics than the relatively big intact bacteriophages and the relatively big antibacterial phage protein molecules. They have better pharmacokinetic properties in human tissues and are less likely to cause undesirable immune reactions. Bacteriophages and antimicrobial phage proteins seem to be valuable instructors in any case in the search for this sort of 'small' antibiotic. Yet the use of bacterial viruses as a cure for bacterial infections dates back to 1921 and was the discovery of the Frenchman Félix d'Herelle. This therapy was however consigned to oblivion, except in the former Eastern bloc, because of the meteoric rise of the antibiotics. There researchers worked eagerly on the further development of phage-based medicines (Vandamme & Raemaeckers, 2003).

Independently of each other, Edward Twort in 1915 and Félix d'Herelle in 1917 discovered the phages and d'Herelle was the first to use the term bacteriophage.

Like all other viruses, phages consist of an outer shell of proteins enveloping a DNA or RNA strand, their genome. Viruses cannot translate or copy their genetic material themselves. They use the "expertise" and "machinery" of other organisms for this purpose. Viruses reproduce at the expense of cells of living organisms. In the case of phages these are bacteria. They are not interested in animal and plant cells. In theory, therefore, bacteriophages seem to be the ideal agent for treating infectious bacterial diseases.

That's what they thought in the former Soviet Union, where research into phage medicine was carried out in earnest. In the period 1920-1950 no fewer than 800 mostly Russian publications on phages appeared. The Eliava Institute in the Georgian capital Tbilisi, which focused entirely on phage therapy, had its heyday between 1970 and 1980. Hundreds of people worked on the production of phage medicines and huge quantities of phages in the form of pills, creams and sprays were sold over the counter. The collapse of the Soviet Union and the ensuing economic crisis also signalled the downfall of the Eliava Institute. It is now trying to survive on the back of a few spin-offs and some phage preparations are still being produced, for instance an artificial skin which is impregnated with viruses to heal skin and burn wounds.

The West was never convinced by the Georgian phage therapy and still isn't. From a scientific perspective, this is not altogether unjustified, because there are still a great many snags in this area, and much more scientific substantiation is needed. Now, however, this reticence seems to be changing, partly due to pressure from the rapidly increasing antibiotic resistance. A Western ode to bacteriophages appeared in the 2007 article "Biotechnological exploitation of bacteriophage research" (Petty, Evans, Fineran, & Salmond, 2007). Aside from the huge possibilities for molecular biology, nanotechnology and the detection of specific bacteria, the authors observed a whole shopping list of potential opportunities for the use of phages to treat infectious bacterial diseases. In short, the phage therapies are no longer on the way out, but are on the way back, as Annet Mooij concluded in Section 9.2 for the "enemy", better known as infectious bacteria. One thing is certain: phage-based antibacterial medicines still have a long way to go in the research centres before we'll find them in (Western) pharmacies.

9.7. CONCLUSIONS

Medical care requires constantly novel antibiotics due to the growing prevalence of resistant pathogens in hospital or community-acquired infections. Notwithstanding this need, major pharmaceutical players seem to be reducing their R&D efforts in the area of new antibiotics. This is due to a combination of factors: considerations about maturity of the new drug candidates, the strong competition among pharmaceutical companies, and the increase in generic antibiotics on the market. There is still a perception that the discovery of novel antibiotics has become a very rare event. Despite significant advances in bacterial genomics, high-throughput screening techniques and synthetic methods, the discovery of novel antibiotics over the past thirty years has not sufficiently kept pace with the demand for new agents. On the other hand, past and present successes suggest a return to microbial products that could be used per se or as scaffolds in the quest for better drugs against multiresistant bacteria. Fortunately, our armamentarium for treating Gram-positive infections is being enriched by novel β-lactams, glycopeptides, lantibiotics and other peptides in different phases of development, and our options are increasing with the introduction of specific vaccines and combinatorial drugs. It has never been more important to understand in detail the mechanisms of, and routes to, resistance in bacteria, so that we can improve the surveillance of emerging mechanisms of resistance to antibiotics introduced in clinics and the environment. We should be aware that emergences and diffusion of bacterial resistance is an unavoidable aspect of evolution, which indeed is closely linked to the magnitude of the selective pressure. This was once more emphasised on the day, 11 August 2010, when we were finishing this chapter by the alarming news in the daily papers that a new Asian superbug has spread from India to the UK. These bacteria have a newly found gene called New Delhi metallo-beta-lactase, or NDM-1, making them highly resistant to almost all antibiotics, including the most powerful class called carbapenems, and experts say there are no new drugs on the horizon to tackle it. As our battle with antibiotic resistance is thus destined to continue, it is of the utmost importance that we learn to use antibiotics cautiously and appropriately. Only in this way can we delay the spread of resistance, a natural phenomenon that will surely not disappear.

9.8. SOURCES

Anonymous. (1995, January). Hiroshima, penicilline en de ENIAC. *PolyTechnisch Tijdschrift*, pp. 36-39.

Barrett, J. F. (2005). Can biotech deliver new antibiotics? *Current Opinion in Microbiology, 8*(5), 498-503.

Burns, M. (2005). *Wartime Research to Post-War Production: Bacinol, Dutch Penicillin, 1940-1950.* University of Sheffield, Sheffield.

Clardy, J., Fischbach, M. A., & Walsh, C. T. (2006). New antibiotics from bacterial natural products. *Nature Biotechnology, 24*(12), 1541-1550.

Cornaglia, G., & Rossolini, G. M. (2009). Forthcoming therapeutic perspectives for infections due to multidrug-resistant Gram-positive pathogens. *Clinical Microbiology and Infection, 15*(3), 218-223.

Cottarel, G., & Wierzbowski, J. (2007). Combination drugs, an emerging option for antibacterial therapy. *Trends in Biotechnology, 25*(12), 547-555.

Dancer, S. J. (2004). How antibiotics can make us sick: the less obvious adverse effects of antimicrobial chemotherapy. *Lancet Infectious Diseases, 4*(10), 611-619.

Demain, A. L., & Elander, R. P. (1999). The beta-lactam antibiotics: past, present, and future. *Antonie Van Leeuwenhoek International Journal of General and Molecular Microbiology, 75*(1-2), 5-19.

Demain, A. L., & Sanchez, S. (2009). Microbial drug discovery: 80 years of progress. *Journal of Antibiotics, 62*(1), 5-16.

Dronda, S. B., & Justribo, M. V. (2007). Will we still have antibiotics tomorrow? *Archivos De Bronconeumologia, 43*(8), 450-459.

Jones, K., Patel, N., Levy, M., Storeygard, A., Balk, D., Gittleman, J., *et al.* (2008). Global trends in emerging infectious diseases. *Nature, 451*(7181), 990-993.

Jovetic, S., de Bresser, L., Tramper, J., & Marinelli, F. (2003). Deacylation of antibiotic A40926 by immobilized Actinoplanes teichomyceticus cells in an internal-loop air-lift bioreactor. *Enzyme and Microbial Technology, 32*(5), 546-552.

Jovetic, S., Zhu, Y., Marcone, G. L., Marinelli, F. & Tramper, J. (2010). ß-lactam and glycopeptide antibiotics - first and last line of defence? *Trends in Biotechnology*, (In Press), Doi: 10.1016/j.tibtech.2010.09.004.

Landsberg, H. (1949). Prelude to the discovery of penicillin. *Isis, 40*(3), 225-227.

Malabarba, A., & Goldstein, B. (2005). Origin, structure, and activity in vitro and in vivo of dalbavancin. *Journal of Antimicrobial Chemotherapy, 55*(Supplement 2), ii15-ii20.

Mooij, A. (2007). De onzichtbare vijand - Over de strijd tegen infectieziekten. Amsterdam, Balans.

Overbye, K. M., & Barrett, J. F. (2005). Antibiotics: where did we go wrong. *Drug Discovery Today, 10*(1), 45-52.

Petty, N. K., Evans, T. J., Fineran, P. C., & Salmond, G. P. C. (2007). Biotechnological exploitation of bacteriophage research. *Trends in Biotechnology, 25*(1), 7-15.

Projan, S. (2004). Phage-inspired antibiotics? *Nature Biotechnology, 22*(2), 167-168.

Shah, D., Dang, M., Hasbun, R., Koo, H., Jiang, Z., DuPont, H., *et al.* (2010). Clostridium difficile infection: update on emerging antibiotic treatment options and antibiotic resistance. *Expert Review of Anti-Infective Therapy, 8*(5), 555-564.

Sipkema, D., Osinga, R., Schatton, W., Mendola, D., Tramper, J., & Wijffels, R. H. (2005). Large-scale production of pharmaceuticals by marine sponges: Sea, cell, or synthesis? *Biotechnology and Bioengineering, 90*(2), 201-222.

Tramper, J., Beeftink, H. H., Janssen, A. E. M., Ooijkaas, L. P., Van Roon, J. L., Strubel, M., *et al.* (2001). Biocatalytic production of semi-synthetic cephalosporins: process technology and integration. In A. Bruggink (Ed.), Synthesis of β-lactam antibiotics - Chemsitry, Biocatalysis and Process Integration. (pp. 207). Dordrecht, Kluwer Academic Pubishers.

van den Berg, M., Albang, R., Albermann, K., Badger, J., Daran, J., Driessen, A., *et al.* (2008). Genome sequencing and analysis of the filamentous fungus Penicillium chrysogenum. *Nature Biotechnology, 26*(10), 1161-1168.

Vandamme, E., & Raemaeckers, P. (2003). Virus komt de mens te hulp. *Natuur, wetenschap en Techniek, 26*(September), 29.

Walsh, C. (2003). Where will new antibiotics come from? *Nature Reviews Microbiology, 1*(1), 65-70.

Williams, P. G. (2009). Panning for chemical gold: marine bacteria as a source of new therapeutics. *Trends in Biotechnology, 27*(1), 45-52.

"We are not doctors and we aren't writing prescriptions for you! We believe that we are smarter than most doctors about steroids. We're sticklers about the truth in anything and we happen to know a lot about steroids (some say that we know too much). This book is telling you what we believe to be practical, real world information incorporating the very latest developments in steroid use. You may not care for our sense of humour, or our attitudes, but we honestly think that there is very little argument in the factual information presented. We happen to be bodybuilders so we do slant the information toward that endeavour. What's important is that most of the drugs we talk about, we've used ourselves a number of times. You should know how a drug really works, not how the label says it's supposed to."

An excerpt from the original *Underground Steroid Handbook*[62]; Daniel Duchaine wrote the book in 1988, just before he went to prison for a year on a steroid charge.

In order for a body to function well, be it that of a human or animal, the many chemical reactions that take place in the cells need to be well regulated and in tune with each other, as do those of the various tissues and organs. Hormones have an important role to play here. They regulate metabolism, reproduction, growth and many other bodily processes. Behaviour and frame of mind are also affected by them.

HORMONE USE CAN HAVE STRANGE EFFECTS

WE BELIEVE WE'RE SMARTER
THAN DOCTORS ABOUT STEROIDS ...

... I EVEN WROTE A
BOOK ABOUT IT!

MY FIRST STEROIDS

[62] www.qfac.com/books/origush.html

10.1. WHAT ARE HORMONES?

Hormones come in all shapes and sizes. Some are steroids, complex chemical compounds, such as our sex hormones (progesterone, testosterone and oestrogen), but bile acids and sterols (e.g. cholesterol)

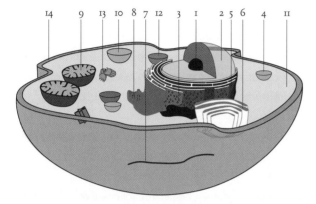

Figure 10.1. Cross-section of an animal cell: (1) nucleolus, a non-membrane bound structure that produces ribosomal RNA; (2) nucleus, cell nucleus containing the DNA; (3) ribosomes, small organelles where protein synthesis occurs; (4) vesicle, a small vacuole, for example, a Golgi vesicle or a membranous vesicle, for transporting larger quantities of material; (5) rough endoplasmatic reticulum, for transporting proteins, with ribosomes on the surface; (6) Golgi apparatus, network in which products like polysaccharides are produced and taken away by budded vesicles; (7) microtubule, cylindrical unbranched tube that fulfils a skeletal function in cells that are not round, for example, nerve cells; (8) smooth endoplasmatic reticulum, carries no ribosomes and is involved in fat metabolism; (9) mitochondria, function in aerobic respiration and generate energy for the cell; (10) peroxisome, microbody where toxic

waste products, such as hydrogen peroxide, are broken down; (11) cytoplasm, living content of a cell, not including the nucleus and large vacuoles; (12) lysosome, an organelle that contains a number of enzymes, whose destructive capacity means that they have to be separated from the rest of the cell; (13) centrioles, organelles that play an important role during nucleus division; (14) membrane, selectively permeable structure, composed mainly of lipids and proteins, which surround cells and also occur within the cells to encase organelles.

also belong to this group. Others are shorter or longer chains of amino acids, peptides and proteins, such as insulin. There are differences between hormones from glands, tissues and cells. The first are produced in the glands and transported via the bloodstream to the organs where they do their work. The tissue hormones only exercise their influence in their close surroundings. Cell hormones regulate all processes in the cell which they inhabit. Hormones are also categorised according to their mechanism of action or chemical structure. They are identified by specific molecules or receptors in the cells of their target organ. The receptors are usually proteins at the cell membrane (e.g. receptors for insulin or adrenalin) or in the cytoplasm (e.g. receptors for oestrogen or progesterone). The hormones regulate by latching onto enzymes or nucleic acids (the building blocks of the genetic material, DNA and RNA). Plants also have such regulators, called phytohormones.

When a hormone is not produced in the correct quantities in an organism, various anomalies occur. Sometimes the changes are a natural process; for instance, a reduction in oestrogen levels during the menopause in aging women or a gradual decline in testosterone levels in the andropause or male menopause in aging men. Where there is a deficiency, it is possible in some cases to top-up hormones using medicines. The first example of this was human insulin made from recombinant bacteria, already described in Chapter 1. In the present chapter three other examples of human hormones are discussed, namely growth hormones (somatotropin or somatrophin), erythropoietin (EPO) and follicle-stimulating hormone (FSH). All three are products of modern biotechnology.

Growth hormone as medicine

Human growth hormone has been firmly placed on the medical map since the end of the 1950s. At that time clinical researchers discovered this hormone in the pituitary, a gland the size of a small pea in the middle of the head under the brain. The pituitary is regarded as the 'master gland' because many hormones from there regulate the excretion of hormones in other glands. In times gone by, the pituitary was called the seat of the soul. Hormone deficiency can be congenital or the result of a cyst, tumour, radiation or trauma. The consequences for children and adults can be far-reaching. For every ten thousand births, one newborn baby suffers from a growth hormone deficiency of some degree of severity. These children grow (much) slower and have (a lot) less muscle mass. Growth hormone deficiency in adults can also have a great many serious consequences.

Following the discovery and development of growth hormone into medicines, pituitary glands from deceased people were for many years the only source from which this hormone could be isolated. The use of the thus acquired somatotropin was strictly regulated and only prescribed in children with a serious form of dwarfism as a result of a lack of self-made growth hormone. In general the children who were treated underwent a surprising recovery in growth and were consequently spared the (psychological) misery of dwarfism. As a rule-of-thumb the treatment was started before puberty and stopped as soon as the epiphyses, the cartilaginous endplates of the bones, had fused (Pownall, 1994).

Tragically, however, it became clear that the therapeutic use of growth hormone acquired in this way also brought with it major risks, e.g. the transfer of diseases. The most horrific of these is Creutzfeldt-Jakob disease (caused by prions and comparable to Mad Cow Disease), which raised serious questions at the beginning of the 1980s about the use of growth hormone extracted from pituitaries. In 1985 the use of this hormone in the Netherlands was discontinued. In an article entitled "*Illegale hormonen*" (illegal hormones) which was published as a scientific editorial in the *NRC* on 18 July 2002, the following appeared. To their knowledge, one of the more than 560 children treated with growth hormone extracted from the pituitary gland of deceased people has died of Creutzfeldt-Jakob disease. Fortunately, the sale of human growth hormone, obtained from recombinant *E. coli* bacteria, has now made the bizarre extraction from human pituitary glands unnecessary. This method was introduced in 1985, shortly after insulin came onto the market as the first hormone product of recombinant DNA technology. The commercial availability of much bigger quantities of safe growth hormone has also opened the way to other applications, not all of which have been equally desirable, as we will see in the following paragraphs.

New research has meantime shown that the fusion of the epiphyses does not signal an end to the need for growth hormone. Adults with growth hormone deficiency often have all kinds of deficiencies, such as an abnormal physical make-up, a poorer physical condition, a different fat metabolism (accumulation of fat tissue), increased cholesterol levels and as a result more cardiovascular disorders, porous bones, sexual disorders, and an overall reduced quality of life and reduced life expectancy, even if they have been treated with growth hormones as children. Clinical studies carried out in the late 1990s convinced the American authorities to approve growth hormone treatment of adults with a deficiency. According to Ken Attie, a clinical researcher who worked at Genentech (a manufacturer of recombinant growth hormone) in San Francisco at the time of his pronouncement, in the US alone there are 70,000 adults with a growth hormone deficiency (Pramik, 1999). This alone, however, is insufficient to explain the growth in demand (Figure 10.2), as predicted at the end of the 20th century.

PRESENT MARKET: € 1.5 BILLION FUTURE MARKET: € 3 BILLION?

Figure 10.2. Expected trend in demand for human growth hormone.

Growth hormone as anti-ageing agent

Children and adults with growth hormone deficiency have relatively large amounts of fat and very little 'lean' tissue mass. Their muscles are poorly developed and they have little stamina, while their kidney function is impaired and their blood pressure is low. This is proof that growth hormone does more than just regulate the growth of a child. When adults with a deficiency receive growth hormone, there are recorded reductions in subcutaneous and abdominal fat of 13 and 30% respectively. More muscle tissue is also developed. The media has exaggerated these types of results and made growth hormone out to be a sort of elixir of youth. However, non-prescribed and unsupervised use is very unwise, because of the potential occurrence of all manner of side effects such as fluid accumulation (oedema), headache and an unhealthy reduction in blood sugar levels (hypoglycaemia), especially at high doses.

In 1990 an article was published on the effects of human growth hormone on men over 60 (Rudman *et al.*, 1990). It suggested that a short course of recombinant growth hormone could reverse ageing symptoms, such as paunch, atrophied muscles, double chin and reduced sexual performance, in otherwise healthy men. The article was based on a six-month study of twelve elderly men ranging from 61 to 81 years of age. As a result of this article rejuvenation anti-ageing clinics offering the human growth hormone (HGH) sprouted up all over the place, particularly in the US, and popular science books and articles were published with enticing titles like:

"Grow young with HGH: the amazing medically proven plan to reverse aging" (Klatz & Kahn, 1998).
"HGH: Age-reversing miracle" (Elkins, 1999).
"Staying young: growth hormone and other natural strategies to reverse the aging process" (Gilbert, Jamie, & Gross, 1999)
"The new anti-aging revolution: stopping the clock for a younger, sexier, happier you" (Klatz & Goldman, 2003).
"Sweet syringe of youth" (Langreth, 2000)[63].

In 2003 Dr. Mary Lee Vance from the Medical department of the University of Virginia Medical Centre, published the article "Can growth hormone prevent

[63] www.forbes.com/forbes/2000/1211/6615218a.html

ageing?" (Vance, 2003). She reported that there were various websites offering growth hormone in oral or aerosol form. In her opinion efficacy has not been shown in any of these substances. She concludes that follow-up studies confirm the findings of Rudman *et al.* concerning changes in physical condition, but improvements in functioning have not been observed. In her view fitness has a more positive effect.

GROWTH HORMONE IS USED FOR ANTI-AGEING

I'M 90 YEARS OLD, BUT STILL LOOKING FANTASTIC

In 2007 a review article on this appeared in the renowned journal published by the American College of Physicians (Liu *et al.*, 2007). The seven authors, all with a medical background, analysed all clinical trials on growth hormone to determine whether it could be used safely and effectively by healthy older patients. They deliberately excluded studies in which the efficacy of growth hormones was evaluated on the treatment of a specific disease. They conducted this research because the use of growth hormone as an anti-ageing agent is very controversial and yet is used as such by many people, even though it has not been approved for that purpose by the FDA, and its dissemination as an anti-aging agent is therefore illegal in the US.

The use of growth hormone as an anti-aging agent scores as one of the highest health-related searches on the internet. According to Mary Lee Vance more than one and a half billion dollars' worth of growth hormone is sold every year, a third of which is probably not under prescription and therefore illegal. Proponents of the use of growth hormone as an anti-aging agent claim that in 2002 more than one hundred thousand people obtained growth hormone without a prescription. Liu *et al.* conclude that there are few published data on the effect of growth hormone on the elderly, but available evidence suggests that the risks far outweigh the benefits if the hormone is used as an anti-ageing agent in healthy elderly patients. The most frequently occurring side effects were oedema in soft tissue and joint pain, while few significant positive effects on physical make-up were reported. In short, there is little chance of growth hormone being made available on the market as a legal anti-ageing agent.

Growth hormone as performance-enhancing drug

The controversial fact that 'fat tissue disappears and muscle tissue appears' when growth hormone is used has also made it irresistible to athletes since the early 1980s and has led to illegal use. This became only too obvious when in January 1998 newspapers reported that the Chinese federation had pulled out a swimmer and coach *en route* to the world championships in Perth. The team members' bags were examined on arrival at Sydney airport during a routine control. The customs officer found a thermos flask containing 26 ampoules in one of the swimmers' bags. Thirteen of them were filled with a liquid. The other thirteen contained a powder described on the label as containing human somatrophin, i.e. human growth hormone. The female owner of the luggage told customs that the ampoules belonged to her coach. The customs department were in no doubt, however, as to their purpose.

A CUSTOMS OFFICER DISCOVERS SOME AMPOULES....

NO SIR, THAT'S WATER FOR MY CONTACTS...
...YOU KNOW... SWIMMING POOL, CHLORINE, EYES...

The online search we conducted as a result soon revealed why they were so convinced. We found a website[64], which is now banned, containing a 9-page manual for the "underground" user and which opened with a similar quote to that at the start of this chapter, also from the Underground Steroid Handbook. The last paragraph of the manual began as follows: "The active substance somatrophin is available as a dry powder and has to be mixed with the solution in the accompanying ampoule before being injected." There

[64] www.bodypage.nl/groeihormonen_of_sth.htm

is a fear, justified by this internet article, that this is just the tip of the iceberg of a widespread use of growth hormone as a performance-enhancing drug. If this explains the phenomenal growth in demand, this wonderful medicine, one of the first products of modern biotechnology, will be seriously but unjustly discredited.

Even more distressing is the example in the previously mentioned article "*Illegale hormonen*". According to this article illegal growth hormone, extracted from the pituitaries of deceased people, was on sale in the Dutch bodybuilders' circuit back then. Bodybuilders who used this product, of Russian origin, run the risk of getting Creutzfeldt-Jakob disease in 10 to 20 years. The Dutch Health Inspectorate therefore issued a direct warning against the use of this illegal growth hormone. In our view it is unwise from a scientific perspective to use recombinant growth hormone without a prescription. So using this sort of illegal pituitary extract is tantamount to playing Russian roulette.

There has been a urine test for tracing human growth hormone since 2004, which meant it was available for the Olympic Games in Athens. The test was developed by the German endocrinologist Prof. Christian Strasburger, who has since improved it. In response, the World Anti-Doping Agency (WADA) announced that athletes would be thoroughly tested for performance enhancement with human growth hormone for the first time at the Olympic Games in Beijing. The science section in the 26/27 July 2008 edition of the Dutch newspaper *NRC*, was largely dedicated to this doping control. Professor of Movement Science and doping expert Harm Kuipers, a former World Champion speed skater, says in this article that the WADA announcement is *window dressing* to a large extent. The test only traces HGH and this disappears from the urine a few days after administration. An "indirect" test on measurable, long-term effects in the body after HGH use is much more necessary (as with EPO, see following paragraph). Looking at the available studies, Kuipers doubts very much whether HGH actually does anything, So he advised the WADA to investigate this first, and then, if the growth hormone does actually make a difference, to invest in an "indirect" test. Recently a review has been written by scientists from the WADA and the IOC (Barroso, Schamasch, & Rabin, 2009). From the same year is the review Growth Hormone in Sport: Beyond Beijing 2008 (Segura *et al.*, 2009).

The importance of this topic is further stressed by the editorial "Game over for sports cheats" written by Vicky Heath, the Chief Editor of *Nature Reviews Endocrinology* (Heath, 2010). She welcomes the "groundbreaking new initiative" that WADA recently rolled out: the athlete biological passport. She writes: "in addition to the usual random drugs tests, athletes are now required to undergo regular monitoring of biological variables, such as their levels of hemoglobin or red blood cell count. Plotting these measurements over time and looking for abnormal variations should facilitate indirect detection of doping because the downstream effects often remain evident long after the actual drug has disappeared from the body. On the face of it, longitudinal screening is an excellent idea, but only time will tell whether the athlete biological passport proves an effective deterrent." Like her we stress that the potential for gene doping must be explored (see also next chapter on gene therapy) and we would like to end this section with an outcry from her, which we also fully endorse: "the use of performance-enhancing drugs is not restricted to elite athletes; published data suggest that abuse of anabolic androgenic steroids and other endocrine drugs is on the rise among high-school and college athletes. Therefore, it is crucial that young people and their mentors are properly educated about the risks involved. Clinicians and other health-care professionals should take the initiative in this respect, as they have a duty of care to highlight the uncertain benefits and potentially harmful effects of doping."

10.3. ERYTHROPOIETIN (EPO)

"Did biotechnology ruin the Tour de France?" So ran the headline of the article by Gaby van Caulil in *BIOnieuws* of 8 August 1998 (Van Caulil, 1998). In it he writes about the "dual" use of erythropoietin, or EPO. EPO is a hormone made by our kidneys, which regulates the production of red blood cells (erythrocytes) in bone marrow. Red blood cells are packed with haemoglobin. This is the pigment that gives blood its red colour and ensures that oxygen is transported from the lungs to other bodily tissue. In 1985 molecular biologists inserted the gene that codes for human erythropoietin into the genetic material of animal cells. These recombinant cells make it possible to produce human EPO on a large

scale for therapeutic treatments in very simple "bioreactors". In 1988 the American company Amgen, based in California, brought recombinant EPO produced in this way onto the market.

EPO is prescribed to patients with anaemia resulting from kidney problems, and to cancer patients receiving chemotherapy. Before recombinant EPO came onto the market, severe forms of anaemia were treated by blood transfusions with all the accompanying limitations and risks (transfer of viral infections such as AIDS or hepatitis B and C, and immune system problems). The availability of recombinant human EPO (r-huEPO) on the market has dramatically improved the treatment of anaemia and it is now one of the world's best-selling medicines. The question is whether it all finds its way to the real patients.

For decades, top sportsmen and women have been trying to boost their levels of red blood cells, because it enables them to take up oxygen more easily. That's the purpose of altitude training. The thin air stimulates the body to produce extra erythrocytes, enabling the sportsmen to perform better when they return to lower altitudes. That EPO is used as a blood doping agent is a known fact and famous top sportsmen and women have admitted taking it. Performance enhancements of approximately ten percent are possible. In top sports this is a monumental improvement. However, injecting EPO is dangerous. The hormone thickens the blood, making it more difficult for the heart to pump this bodily fluid around the body. The sudden death of 18 Dutch and Belgian cyclists between 1987 and 1990 is still shrouded in mystery. These sportsmen died from unexplained cardiac arrest (some of them in their sleep, when the heart rate is at its lowest). There are stories of cyclists having to get up every two hours during the night to use the home trainer in order to keep their circulation going and thus stop their heart from giving out (Bloembergen, 2007).

And yet, EPO as a performance-enhancing drug is still popular among sportsmen and women, because it remains difficult to trace with certainty. An indication of EPO use is the red blood cell count. This is expressed as a percentage of the total blood volume, the haematocrit value. A value of more than 50% suggests EPO use and the person in question is then suspected of doping. The 1998 Tour de France has become notorious because of the great many participants suspected of using EPO, determined on the basis of the haematocrit value. Up until 2000, however, this suspicion was very hard to substantiate.

Fortunately, strenuous efforts were and still are being made to develop more watertight tests. A watertight EPO test will not only protect sportsmen and women from the massive pressure of commerce, but also heal the tarnished reputation of this great medicine.

In the journal *Nature* researchers from the French National Anti-doping Laboratory describe a more effective test (Lasne & de Ceaurriz, 2000). EPO is a protein that consists of 165 interlinked amino acids. The amino acid chains in the recombinant EPO are identical in principle to that of the naturally occurring EPO. Yet there is a difference. These amino acid chains have side chains that consist of saccharide molecules (the chain is glycosylated). Different saccharide chains are linked to the natural EPO than to the recombinant hormone. Taking this difference as a starting point the French researchers designed a detection method (Textbox 10.1). It is, however, a labour-intensive and expensive way of tracing in multiple stages (Van 't Hoog, 2008).

However, during the 2000 Olympic Games in Sydney this test was used, in combination with an indirect blood test which was also published that same year by Australian researchers in the June edition of *Haematologica*. Rinze Benedictus (Benedictus, 2000) called it a "Sherlock Holmes method", whereby a number of different blood parameters are integrated in a model. By entering data from dozens of blood samples into the model, a mathematical representation can be built

ALTITUDE TRAINING...

CYCLING IS FUN... CYCLING IS FUN...
CYCLING IS FUN... CYCLING IS FUN...

The French EPO urine test uses electrophoresis, a chemical separation technique based on differences in electrical charges. Urine samples are placed at the top on a plate of electrophoresis gel (figure 10.3) and a difference in voltage is applied across the plate. In the figure the cathode (negative) is at the top and the anode (positive) at the bottom. A gradient in pH is also applied across the plate; from top to bottom, the pH lowers and it therefore becomes more acidic. Proteins without side chains have positive and negative charges. The more acidic the more positive charges, and the less acidic the more negative. At a certain pH, there are exactly the same number of negative as positive charges and the protein is electrically neutral. This is called the iso-electric point (pI).

Since the recombinant and natural EPO have the same amino acid chain in principle, they can't be differentiated in this way. The side chains, however, consist of different sugar molecules which have a negative charge at a higher pH. This means that for a certain pH (e.g. 5), the balance between negative and positive charges is different for recombinant and natural EPO. The sugar side chains of the natural EPO have relatively more negative charges than the recombinant EPO, in other words, the pI of natural EPO is lower than that of the recombinant EPO, respectively in the range 3.9 - 4.4 and 4.4 - 5.1. EPO samples are brought up to the negative upper side at pH = 5.2, i.e. both natural and recombinant EPO have more negative than positive charges. The effect of the voltage applied across the plate is to move both proteins downwards. When the recombinant EPO reaches pH ~ 4.8 (~ pI), there are as many negative as positive charges, and the molecule ceases to move. That only happens with the natural EPO at pH ~ 4.2. There are several bands visible because natural and recombinant EPO both have several (iso)forms with small reciprocal differences (they are microheterogenic). The bands are made visible using two dyeing methods (double immuno-blotting and chemoluminescence). This 'double blot' procedure makes the test unique and minimises the chance of false positive results (Coons, 2004).

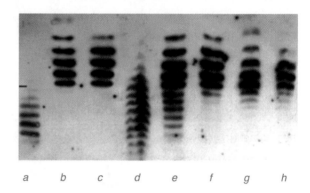

a   b   c   d   e   f   g   h

Figure 10.3. EPO test using electrophoresis (from Lasne and Ceaurritz, 2000; adapted); a. purified from urine, commercially available; natural human EPO; b. commercial recombinant EPO-β; c. commercial recombinant EPO-α; d. urine from a control person; e. and f. urine from two anaemic patients treated with recombinant EPO-β; g. and h. urine from two 'positive' cyclists from the 1998 Tour.

up so that with the right threshold value it selects EPO users and doesn't give false positives. There are two versions of the model: one that predicts use up to three weeks before the test, and one that gives a positive result for doping a few days before. Internist and EPO expert J. Marx, of the University Medical Center Utrecht, said at the time that this theoretically signalled the end of EPO as an undetectable performance-enhancing drug. "But because this indirect, Australian model will probably not stand up to legal scrutiny, the French urine test still has to be carried out after the blood test. Then the sportsman or woman has absolutely no chance", says the EPO expert in the article by Benedictus.

After Sydney 2000 the WADA decided that the French urine test alone was reliable enough to sanction a sportsman or woman caught using EPO (Köhler, 2008). Backed by the IOC, which used to draw up the doping list and approved the doping labs, WADA is the highest doping authority in the world. Thus with the EPO urine test WADA opted for a test that (directly) shows the EPO molecule itself. The results of EPO use are visible in blood, but that is an indirect test. Just as Marx expected, indirect tests are difficult from a legal standpoint. Like Benedictus, Köhler uses detective work as a metaphor: there is a body, and probably a murderer too, but no murder weapon. In 2004 the urine test was validated and every year we hear the same cry yet again: this will be the cleanest Tour ever! So far, however, this has proved to be anything but the case. In 2006 the Tour was a complete disaster in terms of doping scandals

(Textbox 10.2). The 2007 Tour was also seriously tarnished. For the first time in history a leader and wearer of the yellow jersey was disqualified from the course. In 2008 the Tour was hit on the first day and after the first week two "positive" cyclists had already been sent home by their teams. A week later the same occurred for a two-times stage winner. The editor of de Volkskrant ("Any hope of clean Tour is again lost") on 19 July 2008 wrote that in the case of the first two cyclists we could still console ourselves with the thought that neither had a place in the overall ranking. But for the latter case it was a completely different matter:

"Now that one of the most promising young cyclists has succumbed to the temptation, the link between doping and cycling has become all too clear. ....It is as true as it is impotent to conclude that the latest doping affair has brought the credibility of cycling in general and of the Tour in particular to a new low point. Because how is it possible to restore credibility? Stricter controls are only part of the solution, if only because the limits of what is still defensible on legal and human grounds are becoming visible."

At least with a really watertight test the legal possibilities improve. The criticism of the use of the direct urine test alone is gaining momentum, as demonstrated in an interview with the doping expert Rasmus Damsgaard (Randewijk, 2008). There are also increasing calls for a combined direct and indirect test (Köhler, 2008). The above-mentioned Harm Kuipers says in this latter newspaper article that the international skating union has been

This was the headline to an intermezzo in an article entitled "Doping as perpetual motion" by Marije Randewijk in the Dutch Volkskrant of 12 July 2008. It is an interview with Jörg Jaksche, one of the leading players in Operación Puerto:

"On 23 May 2006, following a raid by the Guardia Civil on the laboratory of haematologist Merino Batres and the Madrid apartments of Dr. Eufemiano Fuentes, 220 blood bags, growth hormones, anabolic steroids and EPO were seized. On the same day Fuentes and Manolo Salz, team leader of Liberty Seguros, were arrested. … Three more key figures in the biggest doping affair in the sport were picked up. …The collaboration with Fuentes cost top favourites Ivan Basso and Jan Ullrich their place in the Tour of 2006. Most of the Spanish cyclists had a lucky escape. …The file remains open. …"

One of the statements made by Jaksche in the interview on the subject reads as follows: "The doping problem is not my problem, or that of Basso or Ulrich, it is a problem of the system. We are all victims and perpetrators alike. You are forced to go along with it, and because you do it, you force others to come along with you. It's perpetual motion."

In the first week of October 2008, the Spanish courts dismissed the investigation into this biggest ever doping scandal, thereby permanently closing the Operación Puerto file from a legal point of view. "Only the sport of cycling refuses to close the book on 'Puerto'" according to an article in the NRC of 4 October 2008. "It's better to randomly search for perpetrators than act as if nothing is going on."

combining blood tests with urine tests since 2000. The main problem with the urine test on its own is that there are an increasing number of variants of recombinant EPO on the market, which look more and more like the natural EPO, particularly because human tissue rather than animal cells is now used for production. Unfortunately, the latest developments in the area of modern biotechnology are gratefully used in many illegal laboratories. In the *Korea Times* of 22 February 2006 there was a report on one such development. A team of researchers from the University of Gyeongsang in South Korea announced that they had genetically modified nine mice in such a way that they produced EPO in their milk. According to the researchers the levels in the milk were so high and stable that they could be used in clinical experiments. The intention is, amongst other things, to incorporate the EPO in medicines for patients with cardiovascular disorders. Its use in the sporting world is obviously not mentioned.

In order to increase production still further a number of pigs have also now been modified and the hope is to obtain more EPO from pig milk, if possible at the lowest possible cost price. Whatever happens, it is virtually impossible to trace all new types that come on the black market. According to Damsgaard the only solution is to develop a benchmark of what a normal test result should be for an individual sportsman or woman.

But is a watertight test the end of the story? Probably not. Researchers who have inserted the gene for EPO in mouse muscles with the intention of developing therapeutic applications, are afraid that once their research delivers results, sportsmen and women will also use this gene therapy for doping (see also Chapter 11 on gene therapy). The ethical question is whether this is reason enough to halt such research, which could be life-saving for some.

On 3 July 2010 Jamey Keaton, Associate Press Writer published the article "Doping lurks as wild card at 2010 Tour."[65] He writes: "Last year's Tour de France was notable for more than just Lance Armstrong's return. It was also free of positive doping tests. This came after three straight years during which cycling's main event was marred by drug cheats. But those who believe this is a sign that the drug-plagued sport is turning a corner should think again. At a race for cycling's budding stars several weeks after the 2009 Tour, no positive doping tests turned up either—until customs officials raided a Ukrainian team bus, seized doping gear, and investigators later wrested admissions of blood doping and use of endurance booster EPO during the event. The 2010 Tour began on Saturday 3 July in Rotterdam, with cycling bosses holding their breath in the hope that the race would be clean—and not just in appearance." Now three weeks later the Tour has a winner again, the same as last year, i.e. the Spaniard Alberto Contador. Will it again be a Tour free of positive doping tests? For that we have to wait another couple of weeks for the results, too late probably for this book. But if so, does that mean then that Pandora's hope is lifted from the bottom of the box, or does it herald the new era of gene doping? Let's hope not!

In conclusion, as far as EPO as a medicine is concerned, there may well be more areas of application in the future. In 2003 German researchers came to the conclusion that EPO is also a good candidate for the treatment of schizophrenia. The hormone would not so much tackle the symptoms (paranoia and delusions) as the underlying neurodegeneration, which proceeds despite successful medication. A clinical trial has begun in eight German centres. The same researchers had already observed the beneficial effects of EPO in 2002 in the repair of a brain haemorrhage. The status of this research can be found on the website of the group.[66]

---

[65] sports.yahoo.com/sc/news?slug=ap-tourdefrance-doping

10.4. PUREGON™: FOLLICLE-STIMULATING HORMONE (FSH)

Sooner or later in their lives most adults want to have children. Those who fail in their attempts to fulfil this desire often feel frustrated, desperate and hopeless, with all the ensuing consequences. In our Western society about 15% of all couples have fertility problems. Fertility hormones have been prescribed for decades to solve this issue. The prescription of the fertility hormone FSH made from genetically modified animal cells is fairly new. These cells are originally isolated from the ovary of a Chinese hamster and can then be grown in flasks and bioreactors.

In 1923 Organon, once part of Akzo Nobel but since 10 March 2009 part of the American pharmaceutical company Merck, began producing insulin under the licence of the University of Toronto. *Zwanenberg Vleesbedrijven*, a meat company, supplied its subsidiary Organon with "worthless" slaughter waste - in this case pancreas - from which Organon isolated the valuable insulin. Less than a decade later, this approach still appeared to be a golden formula; in 1932 Organon began with the isolation of the valuable fertility hormone hCG (human Chorionic Gonadotrophin) from the "waste" urine of pregnant women. The hormone which induces ovulation was sold under the brand name Pregnyl® and was used in fertility treatment. This would later be the scenario for the waste urine of postmenopausal women. This appears to be a rich source of two other members of the family of fertility hormones: follicle-stimulating hormone (FSH) and luteinising hormone (LH). These are the active components of the preparation Humegon® which stimulates follicle ripening.

Thanks in part to the advent of *in-vitro* fertilisation (IVF), in which both the above-mentioned preparations are used, the collection of urine samples in the Netherlands and beyond has risen to millions of litres per year. The collection of this special urine has always been a delicate matter because it requires the cooperation of volunteers. In less than 80 years this urine collection has undergone a veritable metamorphosis. First it was collected by bike in the town of Oss. Then a regional collection was conducted by horse and cart. After the Second World War the operation was expanded to the whole country by car. The organisation '*Moeders voor Moeders*' (Mothers for Mothers) now has a whole fleet of ten tonne trucks for the job.[67]

THE COLLECTION OF URINE FROM PREGNANT WOMEN IS VERY SUCCESSFUL!

USE ONE OF THE TOILET BOOTHS, MADAM...

NO DRINKING WATER!

---

[66] www.neuroprotection-schizophrenia.de

[67] www.moedersvoormoeders.nl

Fertility hormones are extracted from urine using classical biochemical methods. The foremost complications are the variable quality of the urine, the large scale of production and the strict purity criteria which the injection preparations have to satisfy. Reasons enough for Organon to conduct a search in the 1980s for an alternative production method using modern biotechnology. In July 1996 they hit the jackpot. Organon came on the market with human FSH (Puregon™), made with genetically modified Chinese hamster cells.

FSH is one of the most complex protein molecules in the human body. It consists of two protein chains and is encoded not by one but by two genes. Saccharide groups then have to be linked to the protein chains (a process called glycosylation) if the FSH is to be activated (Figure 10.4). Genetically modified microorganisms seem unable to glycosylate. In contrast, after genetic modification, Chinese hamster cells seemed perfectly capable of performing the whole process; moreover, they are safe and widely accepted for the preparation of recombinant medicines.

Since Puregon is more than 99% pure, it can be administered to patients subcutaneously rather than intramuscularly. Patients can also inject themselves.

Follicle Stimulating Hormone (FSH) is composed of two protein chains (α and β) and contains a number of sugar groups. For the production of FSH the Chinese Hamster Ovary (CHO) cell line is used.

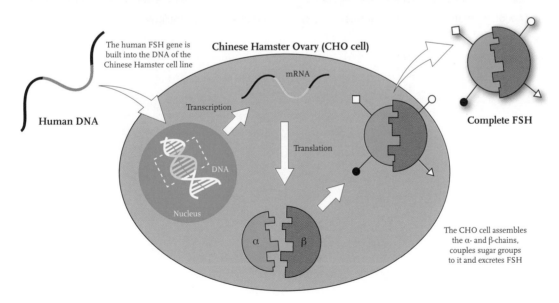

Figure 10.4. Mammalian cells for FSH production, adapted from Olijve & Houwink (1993).

The recombinant FSH is homogenous and contains no human protein impurities, as is the case with traditional preparations. This makes it ideal for use in patients who are allergic to FSH from urine and thus have to cease further treatment. Because the manufacture of Puregon does not rely on the availability of the raw ingredient (urine), production is more flexible. In short, a great product, a fantastic production process, and fortunately no incorrect applications (at least as far as we know). So has the organisation "Mothers for Mothers" become obsolete? Not quite yet. As long as there are no better alternatives for the other fertility hormones that are also obtained from urine, this organisation will still play a vital role. However, it is highly likely that sooner or later modern biotechnology will come up with alternatives for these also. In fact it has almost reached that point. Both the two other mentioned fertility hormones are available in recombinant form and are approaching practical application.[68, 69, 70]

10.5. IN CONCLUSION

In this chapter we have taken a close look at three hormones made with the help of modern biotechnology and sold as medicines. Two of them save lives and the third creates lives. All three are really wonderful, pure medicines with many benefits compared to the old drugs. Yet two of the three have a tainted reputation, mainly because of their illegal use as performance-enhancing drugs. That's a crying shame, especially since it is precisely these two that are often used as life-saving treatments in patients. Nevertheless, they are still beautiful examples of very worthwhile products of modern biotechnology.

[68] www2.cochrane.org/reviews/en/ab005070.html
[69] www.gfmer.ch/Endo/PGC_network/Recombinant_luteinizing_hormone_Pou.htm
[70] www.fertstert.org/article/S0015-0282(09)00502-0/abstract

10.6. SOURCES

Barroso, O., Schamasch, P., & Rabin, O. (2009). Detection of GH abuse in sport: Past, present and future. *Growth Hormone & LGF Research, 19*(4), 369-374.

Benedictus, R. (2000, 16 September). EPO-doping in toom. *Bionieuws*.

Bloembergen, J. (2007, 7/8 July). Limonade in de benen. *NRC*.

Coons, R. (2004, 1 November). New detection methods help level the field. *Genetic Engineering News*.

Elkins, R. (1999). *HGH: Age-Reversing Miracle*. Orem, UT, Woodland Publishing.

Gilbert, E. M. D., Jamie, J.J., & Gross, S. (1999). *Staying young: growth hormone and other natural strategies to reverse the aging process*. Boca Raton, FL, Age Reversal Press.

Heath, V. (2010). Game over for sports cheats? *Nature Reviews Endocrinology, 6*(8), 413.

Klatz, R., & Goldman, R. (2003). *The new anti-aging revolution: stopping the clock for a younger, sexier, happier you!* (3rd ed.). Laguna Beach, CA, Basic Health Publications, Inc.

Klatz, R., & Kahn, C. (1998). *Grow Young with HGH: The Amazing Medically Proven Plan to Reverse Aging*. New York, Collins.

Köhler, W. (2008, 26/27 July). Epodope. *NRC*.

Langreth, R. (2000). *Sweet syringe of youth*. Forbes Global.

Lasne, F., & de Ceaurriz, J. (2000). Recombinant erythropoietin in urine. *Nature, 405*(6787), 635.

Liu, H., Bravata, D. M., Olkin, I., Nayak, S., Roberts, B., Garber, A. M., *et al.* (2007). Systematic review: The safety and efficacy of growth hormone in the healthy elderly. *Annals of Internal Medicine, 146*(2), 104-115.

Olijve, W., & Houwink, E. H. (1993, October). Biotechnologie vernieuwt geneesmiddelen. *Chemisch Magazine*.

Pownall, M. (1994). Biotechnology triumph brings hope to more patients. *International Biotechnology Laboratory* (May).

Pramik, M. J. (1999, 1 January). Recombinant human growth hormone: one of biotech's first products expands indications. *Genetic Engineering News*.

Randewijk, M. (2008, 19 July). Dopingexpert juicht epo-Tour toe; interview met Rasmus Damsgaard. *De Volkskrant*.

Rudman, D., Feller, A. G., Nagraj, H. S., Gergans, G. A., Lalitha, P. Y., Goldberg, A. F., *et al.* (1990). Effects of human growth-hormone in men over 60 years old. *New England Journal of Medicine, 323*(1), 1-6.

Segura, J., Gutiérrez-Gallego, R., Ventura, R., Pascual, J., Bosch, J., Such-Sanmartín, G., *et al.* (2009). Growth hormone in sport: beyond Beijing 2008. *Therapeutic Drug Monitoring, 31*(1), 3-13.

Van 't Hoog, A. (2008, 16 August). Scheidsrechter met kolom en massaspectrum. *C2W*.

Van Caulil, G. (1998, 8 August). Heeft biotechnologie de Tour de France bedorven? *BIOnieuws*.

Vance, M. L. (2003). Retrospective - Can growth hormone prevent aging? *New England Journal of Medicine, 348*(9), 779-780.

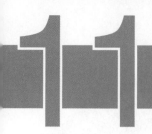

"I just bumped into a man who we admitted to the Antonie van Leeuwenhoek hospital four years ago. I can attribute the fact that he is now in good health due to the gene therapy he received back then."

Winald Gerritsen, director of the Cancer Centre in the Free University Amsterdam Medical Centre, March 1999.

The January 1996 issue of *Chemisch Magazine* contained an article in the New Technological Trends of the 21st century section entitled "Gene therapy causes fourth medical revolution." The article begins as follows:

"More than 4,000 diseases and abnormalities are caused by a defect in a single gene. Although still in its infancy, gene therapy may be the solution. Insiders believe that this therapy will result in a new revolution in the medical world in the 21st century. History has already witnessed three major turnarounds in the fight against disease. The first was with a greater focus on sanitation facilities and sewers, which went a long way to suppressing infectious diseases." - We have already read about Louis Pasteur's pioneering activities in the area of hygiene in the chapter on wine. – "Then came anaesthesia, which enabled doctors to treat patients under sedation. Finally, there was the introduction of vaccines and antibiotics, which prevented and treated many viral and bacterial infectious diseases." Now, more than a decade later, we will address the following question in this chapter: Does gene therapy really herald the fourth revolution? First, however, we will look at what gene therapy actually involves.

THE FOUR MEDICAL REVOLUTIONS

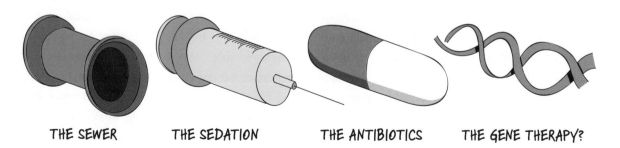

THE SEWER   THE SEDATION   THE ANTIBIOTICS   THE GENE THERAPY?

11.1. WHAT IS GENE THERAPY?

The idea behind gene therapy is simple. When a child is born with a serious disease caused by a gene defect, it is possible in theory to provide the diseased cells with a normal copy of the gene, thereby removing the defect. The practice, however, is far from simple. Whilst it is true that medical and biotechnological researchers have developed a few techniques to insert 'corrective genes' in (diseased) cells, stable integration of a 'working corrective gene' in the DNA of the cells is another matter entirely.

The simplest method is to inject naked DNA, containing the corrective gene, into tissue. Nowadays a DNA molecule is easy to make and to buy. However, in contrast to other cell types, only muscle cells can take up loose DNA. A more efficient procedure is to use a so-called vector, such as a liposome or virus. A liposome is an artificially synthesised bubble membrane containing, for instance, an aqueous solution with DNA (Figure 11.1). The bubble membrane can be made up of a double layer of molecules that resemble phospholipids, the natural molecules that are the main component of the cell membrane in living organisms. The molecules have a hydrophilic (water-loving) head and a hydrophobic (water-repellent) lipid tail. In theory, liposomes can be used to take their content into individual cells by allowing the membrane to fuse with the cell membrane. If the content were to be, for example, the corrective DNA, this could be transferred via the liposomes into the target cell.

Figure 11.1. Liposome with corrective DNA.

The most efficient method is the use of viruses as a vector. The advantage of viruses is that they have a natural tendency to stick to host cells and are able to inject their genetic material into the host cells. This is their way of multiplying and "surviving". The virus DNA then becomes integrated in the cell DNA, causing the cell to make large quantities of the new virus and become "sick". Viruses are the Trojan horses of biology: they are experts at incorporating foreign DNA.

So a virus is a perfected instrument for introducing genes into cells. With the help of recombinant DNA techniques it is possible to replace the genes of a virus that are crucial for replication of the virus with human genes. The virus is consequently weakened so much that it can no longer kill our cells, but is still able to deliver

DNA to the cell. Such viruses are obviously first tested in animals to highlight any of the possible major problems:

- *It is difficult to infect a large number of body cells simultaneously with weakened viruses.*
- *If the infection succeeds, the genes often fail to reach the cell nucleus which contains the cell's DNA.*
- *If they do reach it, they become incorporated in abnormal locations in the DNA, often causing them to be ineffective.*
- *If, in the beginning, the genes appear to be working reasonably well, this action usually stops after a few weeks.*

Despite the disappointing results in animals, medical professionals have still been relatively quick to test gene therapy on patients. Gene therapy has been used, for instance, as the last resort in the case of children with serious and untreatable congenital disorders.

So that it can be used in gene therapy, a corrective gene (or genes) must be incorporated in the virus DNA in such a way that the viruses can no longer replicate in a host cell and so can no longer make this cell sick. Many viruses infect a variety of cell types, while gene therapy is directed at specific cells/tissues. As of this moment, there are two principal ways of using virus vectors for gene therapy to reach this goal. The most common technique involves a doctor removing cells containing the defective gene from the patient's body, adding a corrective gene to these cells in the laboratory using a genetically modified virus and then implanting the cells in the patient's body again (Figure 11.2).

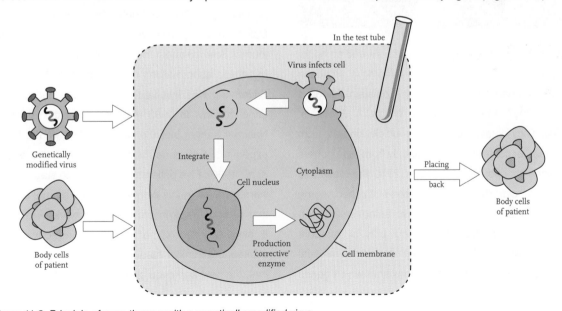

Figure 11.2. Principle of gene therapy with a genetically modified virus.

Initially it was mainly blood cells that were 'corrected', but these have a limited life span and so 'the cure' is also of short duration. As a result current research now focuses on bone marrow cells, which continue to divide and ensure the creation of blood cells.

In the second technique doctors directly administer the virus vector with the corrective gene to the tissue manifesting the gene deficiency. This approach, therefore, is only effective against local abnormalities, such as cystic fibrosis, or inherited muscular disorders like Duchenne muscular dystrophy. In principle, tumours can also be treated in this way by administering a virus vector with a "suicide" gene; then, when the tumour cell is treated with certain chemotherapies, this gene ensures that the tumour cell self-destructs.

11.2. A SHORT HISTORY OF GENE THERAPY

Since 1997 several medical researchers and an editor of a medical journal have been keeping a database on clinical trials for gene therapy (Edelstein, Abedi, & Wixon, 2007; Edelstein, Abedi, Wixon, & Edelstein, 2004). They obtain the data from official bodies, from literature, at conferences and directly from researchers or research sponsors. By June 2010 the database contained almost 1,650 trials, some finished, some ongoing and a few with "start-up authorisation". An analysis is made continuously not only of numbers and geographical distribution but also of the medical reasons behind the trials and the way in which genes were transferred. Details can be found on their very informative and user-friendly website[71]. In June 2010, for example, it showed that the US (64.3%) and Europe (29.3%) had carried out the greatest number of trials by far; the Netherlands came in at 1.6% with 27 trials. Cancer treatment outscored the rest with 64.5%. The great majority (60.5%) of all clinical trials were still in Phase I (see Textbox 9.1. Phases of drug development). In short, this is a very rich source of up-to-date information about gene therapy. Also a rich source of rather up-to-date information is the website[72] of the Human Genome Program sponsored by the US Department of Energy Office of Science, Office of Biological & Environmental Research.

The concept of gene therapy has existed for quite a while. In 1972 Professor Theodore Friedmann and his colleague Richard Roblin, from the University of California in La Jolla, wrote in the leading scientific journal *Science* about the possibilities of gene therapy in genetic abnormalities (Friedmann & Roblin, 1972); Friedmann was involved from the very beginning of gene therapy. In 1989 researchers at the American National Cancer Institute in Bethesda were the first to experiment with gene therapy in humans (Rosenberg *et al.*, 1990). Five patients with advanced stage cancer were involved in this pioneering trial. The trial showed that gene therapy in principle also works in people. It also revealed some important prerequisites for subsequent clinical studies with gene therapy. After that, the next approved clinical trial took place in 1990

---

[71] www.wiley.co.uk/genmed/clinical
[72] www.ornl.gov/sci/techresources/Human_Genome/medicine/genetherapy.shtml

(Blaese *et al.*, 1995). For the first time some success was achieved with gene therapy carried out in two girls with SCID (Severe Combined Immune Deficiency) - a very serious congenital syndrome in which the immune system doesn't work; however, Section 11.3 describes how, ten years later, the treatment of SCID with gene therapy had disastrous results. From 1990 to 1999 the number of clinical trials rose dramatically (Figure 11.3). In general, expectations ran high in this period, but there were still voices of concern to be heard about the possible risks of gene therapy, and some critics pointed out that this treatment had thus far delivered little in the way of therapeutic benefits.

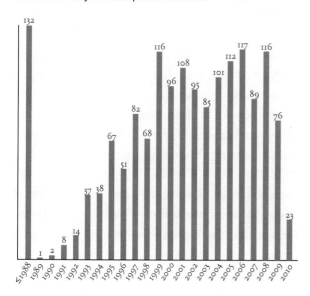

Figure 11.3. Number of approved clinical trials over the years (last updated October 2010[73]).

[73] www.wiley.co.uk/genmed/clinical/

The mood changed, however, at the end of the 1990s following two serious events. The first took place on 17 September 1999 when the 18-year old Jesse Gelsinger died as a result of a gene therapy treatment (Raper *et al.*, 2003). The death was attributed to a totally unexpected and devastating inflammatory response to the adenovirus used (Textbox 11.1). The FDA therefore stopped this and a few other trials. In February 2005 the American Ministry of Justice delivered its final verdict on this case (Couzin & Kaiser, 2005). The University of Pennsylvania, where the trial had been carried out, was held responsible and had to pay a settlement fee of $517,000. In addition, extra restrictions on gene therapy research were imposed on the doctors who had performed the therapy.

In 2000 morale was boosted somewhat by a report from France about successful gene therapy treatment in ten children with a rare form of SCID (Cavazzana-Calvo *et al.*, 2000). However, the joyous enthusiasm didn't last longer than two years, when at the end of 2002 there was an alarming report that the two children in which the greatest number of "corrective" cells had been implanted had developed leukaemia as a side effect. The study was voluntarily stopped and, once the protocol was revised, it was restarted with lower doses of corrective cells. In Section 11.3 we elaborate on this issue.

The death of Jesse Gelsinger and the serious consequences thereof overshadowed a positive result of gene therapy reported at the time. At the end of 1999 the company Avigen declared that the first patients

A virus is a packaged piece of DNA, that specialises in offering up its gene package to host cells. For example, the adenovirus, which is responsible for some of our colds, can bind to almost any cell type and then insert its DNA into the host cells. One such adenovirus, genetically modified with a view to gene therapy, appeared to work well in animal tests, but in practice turned out to be unsuitable for gene therapy. The DNA of this virus is not built into our chromosomes. It therefore only cures as long as the virus multiplies in our cells and that usually stops after a few weeks have passed. Additionally, there is also an immunological complication. People usually have antibodies protecting them against adenoviruses and therefore it is necessary to inject huge quantities of the virus. That is a risky practice in patients with a compromised immune system. Moreover, the weakened virus is often still able to generate a powerful resistance response. This resistance destroys the cells that are infected with the virus and it is precisely these cells that should set the gene therapy in motion. However, gene therapy was and still is a spectacular, new, experimental therapy and successes in this area are bound to attract attention. Such it was that Jessie Gelsinger, who when he turned eighteen had a serious, but not life-threatening metabolic disease, was injected with an enormous quantity of modified adenovirus with the aim of correcting his metabolic disease. Four days later, Jessie died of a massive untreatable immune response to the weakened adenovirus. In one of his columns from 2001 about gene therapy, Emeritus Professor Piet Borst, scientist, columnist and long-time director of the Dutch Cancer Institute in Amsterdam, argued that the gene therapy performed on Jessie Gelsinger was medically and scientifically irresponsible. He concluded (and we totally agree with him on this point): "This is not how it should be done!" After this unfortunate incident the rules in America have obviously been tightened.

---

with haemophilia B were experiencing positive effects of their experimental gene therapy. Haemophilia B is a rare blood-clotting disorder. Because of a congenital gene defect the bodies of sufferers from this disease don't make any coagulation factor IX. Regular injections with coagulation factor IX from donor blood prevents them becoming handicapped or dying of internal (mainly in the joints) or external haemorrhaging. In the 1980s many of these patients contracted AIDS, because their coagulation factor preparation was taken from the blood of seropositive donors. At the time of Avigen's report, the first three patients, who by the end of 1999 had already received experimental gene therapy some months before, were making factor IX themselves again, albeit in low concentrations. Avigen inserted the gene for factor IX in an adeno-associated virus (AAV) (Figure 11.4). Encouraged by the positive results, Avigen is sponsoring clinical trials at Stanford University Medical Center and the paediatric hospital in Philadelphia (Sedlak, 2003).

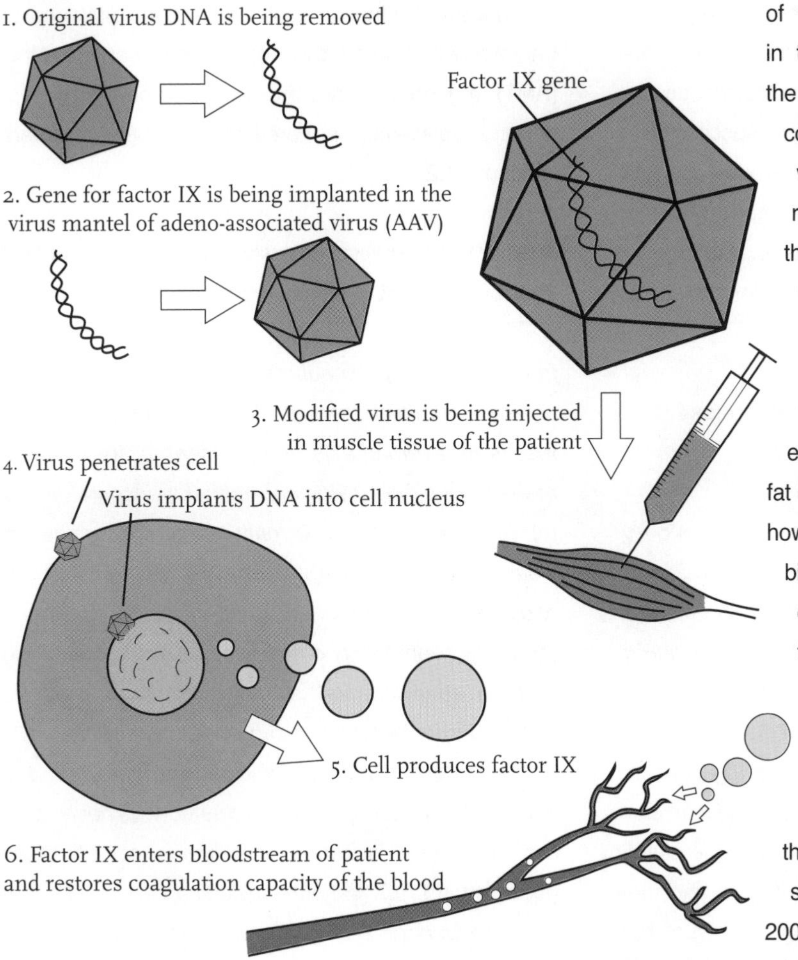

1. Original virus DNA is being removed

2. Gene for factor IX is being implanted in the virus mantel of adeno-associated virus (AAV)

Factor IX gene

3. Modified virus is being injected in muscle tissue of the patient

4. Virus penetrates cell

Virus implants DNA into cell nucleus

5. Cell produces factor IX

6. Factor IX enters bloodstream of patient and restores coagulation capacity of the blood

Figure 11.4. Gene therapy to treat haemophilia B, adapted from Köhler (2000).

AAV is like a cold virus, but without the adverse effects. The immune reaction is minimal and the location where the gene is incorporated is well-known. The risk of it activating a cancer gene is also slight. John Kastelein, a professor of vascular medicine and a specialist in the genetics of lipid metabolism at the University of Amsterdam and his colleague Erik Stroes, investigated whether it was possible to cure a rare lipid metabolic disorder with gene therapy using AAV. About 30 people in the Netherlands have a genetic defect whereby lipoprotein-lipase (LPL) is poorly made or not made at all in their bodies. LPL is a fat-processing enzyme that splits fatty acids from the fat in the lipoprotein. In these 30 people, however, the lipoproteins in the form of big fat balls (chylomicrons) continue to circulate in the blood. They end up in the pancreas, where they can cause a very painful infection. A strict fat-free and alcohol-free diet is currently the only thing that helps a little. Kastelein and Stroes incorporated the LPL gene into an AAV and, following successful animal trials, they began in 2005 with clinical trials on patients. The trial involved eight patients, each of whom had suffered from a pancreatic infection more than five times. In June 2007 Stroes was able to present the first hopeful, provisional results in Seattle at the annual congress of gene therapists.

He reported that the pain after a great many injections under sedation into both thighs was minimal, as were the side effects. The fat content in the blood fell and the effect appeared to be long-lasting. Some of the patients also reported having fewer stomach aches. This first trial was to demonstrate safety. Now the correct dosage must be found. At the very least this early success gives hope to around 5,000 of these patients worldwide; but it may also be important for other, much more frequently occurring disorders in which lipid metabolism plays a role, for example, cardiovascular disorders, diabetes and obesity. Professor Kastelein was in 1998 one of the founders of Amsterdam Molecular Therapeutics Inc. (AMT), a gene therapy company based on the concept of gene replacement in hereditary lipoprotein disorders (Textbox 11.2).

Another provisional success story was published in April 2008 by a British/US research group in the online edition

---

TEXTBOX 11.2.

Glybera: gene therapy for lipoprotein-lipase deficient patients[74].

Lipoprotein-lipase-deficiency (LPLD) is a seriously debilitating, and potentially lethal, orphan disease, for which no approved therapy exists today. The disease is caused by mutations in the LPL gene, resulting in highly decreased or absent lipoprotein-lipase (LPL) activity in patients. LPL activity is needed in order to break down chylomicrons, large fat-carrying particles that are formed in the gut and enter the circulation after each meal. When such particles are not adequately broken down they accumulate in the blood, and they may obstruct small blood vessels, which in turn can lead to pancreatitis. Recurrent pancreatitis in LPLD patients can result in difficult-to-treat diabetes. On June 4, 2010 Amsterdam Molecular Therapeutics reported new data showing that its lead product Glybera results in the break-down of chylomicrons in LPLD patients. Glybera is a gene therapy product that induces functional lipoprotein activity. New data

from an ongoing Canadian clinical study indicate that a single administration of Glybera in LPLD patients results in a remarkable improvement in the ability to break down the chylomicrons that transport dietary fat (triglycerides). LPLD patients are incapable of clearing chylomicrons which are responsible for causing significant morbidity and mortality. "The long-term improvement in chylomicron handling following Glybera administration is very impressive", said Dr. André Carpentier, co-investigator from the University of Sherbrooke, Quebec, Canada, who designed and analysed the chylomicron sub-study. "These data are important, because the major complications observed in LPLD patients, including pancreatitis, are a consequence of chylomicron overload. They also constitute evidence for a long-term clinically relevant lipoprotein-lipase activity induced by Glybera" noted the principal investigator, Prof. Daniel Gaudet, from the University of Montreal, and ECOGENE-21 clinical study center, Chicoutimi, Quebec, Canada. These new data provide a basis for explaining the mechanism of action of Glybera in LPLD patients, and in general for continued pharmacologic activity after one-time gene therapy.

[74] www.amtbiopharma.com

of *The New England Journal of Medicine* (Bainbridge *et al.*, 2008). Six patients between 17 and 26 years old, who had gone virtually blind because of a rare congenital disease, regained a little of their sight following gene therapy. The improvement in the 18-year old British man, Steven Howarth, was particularly spectacular. Here too, the trial was carried out first to demonstrate safety. The researchers now want to treat children whose sight has not deteriorated as much. Trials in young dogs have actually demonstrated a much more drastic improvement in sight than in adults treated thus far.

On 3 June 2010 scientists from Mount Sinai School of Medicine in New York reported to the press (United Press International) that they had developed a gene therapy that is safe and effective in reversing advanced heart failure. The researchers said the therapy, called Mydicar, is designed to stimulate production of an enzyme that enables the failing heart to pump more effectively. In a Phase II study, injection of the gene SERCA2a through a routine, minimally invasive cardiac catheterisation was safe and showed clinical benefit in treating and decreasing the severity of heart failure. The data indicate that SERCA2a is a promising option for patients with heart failure.

A great many clinical trials involving gene therapy are at an advanced stage, but as far as we know there has only been one commercial treatment, and that was in China. In the September 2005 issue of the American journal *Human Gene Therapy* Chinese researchers describe the world's first commercial gene therapy treatment (Peng, 2005). It involved a recombinant adenovirus with a human gene that suppresses certain types of head and neck cancer. This product, Gendicine, was approved by the Chinese authorities on 16 October 2003, after more than five years of clinical trials. The article describes not only the activities that led to the successful market entry of Gendicine, but also the educational campaign to inform the public, the building of the production facility and the technology and quality controls used to guarantee the production of a safe and effective product. The Chinese government's policy of heavily promoting R&D in the area of gene therapy is also emphasised in the article.

11.3. SCID CHILDREN

In 1990 the first partially successful gene therapy procedure was carried out in two girls with a specific form of SCID (Blaese *et al.*, 1995). SCID is an abbreviation for Severe Combined Immune Deficiency. The syndrome consists of several very serious congenital disorders, resulting in a very fragile immune system. Without treatment, patients often die before they reach the age of one or two. In these patients the enzyme adenosine deaminase (ADA) was missing, or not functioning or poorly functioning.

The above-mentioned columnist Piet Borst (Textbox 11.1) wrote that after the Gelsinger drama there was light at the end of the tunnel after all and referred to the group under Alain Fischer in Paris. This concerned the treatment of a rare form of SCID, in which there was a defective gene on the X-chromosome (SCID-X1). As a result of this gene defect in SCID-X1 patients, precursor cells of the immune system do not develop into adult resistance cells. The patients therefore have no resistance to infectious diseases. In people with a properly functioning immune system, intruders are rendered harmless by specialised white blood cells (T cells). Precursor cells of white blood cells usually develop under the influence of growth factors into effective white blood cells, but not so in SCID-X1 patients. These children were therefore not resistant to all kinds of germs that are easily fought off by healthy people. They used to have to live in a 'bubble' to protect them from infection. Now they are administered with the missing resistance cells by means of regular bone marrow transplants.

In 1999 the French researchers "harvested" precursor cells from SCID babies with a bone marrow puncture. They infected these precursor cells with a 'cripple' virus, in which the researchers had inserted the 'corrective' gene. After three days of being exposed to this virus in a test-tube, the precursor cells were put back in the body of the babies (Figure 11.2). After three months of isolated nursing care, adult resistance cells were circulating in the babies' blood. Since they now had a properly functioning immune system, they were allowed to go home. On 21 January 2001, Borst wrote: "In two of the babies the gene therapy took place more than a year ago, and they are still in great condition, according to Fischer. Fischer's success is based on careful preliminary research on cells in laboratory animals, so that he could find out precisely what the best method was of carrying out gene therapy." Two years later, however, it seemed that it was still not enough, when the researchers published that two of the patients treated had contracted leukaemia (Hacein-Bey-Abina

et al., 2003). Initially the researchers thought that the leukaemia occurred because the gene was inserted so unfortunately in the DNA as to accidentally activate an oncogene in the T-cells. An oncogene is a gene that can convert cells into tumours. Four years later, however, a number of American and German researchers reported in Nature of 27 April 2006 that the inserted gene itself was the cause of the disease, and they concluded that far more time was required in gene therapy research to carry out experiments with laboratory animals, before testing on humans.

After the discovery of leukaemia, the research was voluntarily discontinued and only continued with lower doses of corrective cells after the protocol had been revised. In January 2005, however, traces of cell growth were discovered in a third child, this time as a result of another oncogene. This child and one of the first two victims responded well to chemotherapy. Sadly, the other of the two did not, and died in October 2004. Edelstein et al. reported in 2007 that all the other patients in the French study had to date benefited from the gene therapy. On 31 January 2009 it was reported that eight of the ten had been able to live four years on average without medication, according to an international study under the supervision of the gene therapy centre in Milan.

The latest news we have about the "bubble boy" treatment is from 21 July 2010. Gene Emery writes for Reuters that the 10-year study of nine boys born without the ability to ward off germs has found that gene therapy is an effective long-term treatment, but it comes at a price: four of them developed leukaemia. He quotes the team leader of the Paris research group that reported their conclusion in *the New England Journal of Medicine*: "All children except one, including the three survivors of T-cell acute leukaemia, could live normally in a non-protected environment and cope with microorganisms without harmful consequences while growing normally." Earlier that month, on 7 July 2010, Amy Dockser Marcus wrote in The Wall Street Journal that researchers have launched a new gene-therapy trial for children with the rare disease known as "bubble boy syndrome". In the new study scientists plan to enrol 20 boys with SCID-X1 at five sites around the world. In this new trial researchers will take stem cells from a patient's own bone marrow, deliver a functioning gene into those cells in the lab and then infuse them back into the patient. The researchers believe they have stripped out the feature of the treatment that caused leukaemia. The parts of the vector thought to have activated leukaemia-causing genes have been taken out. Study participants will be monitored for 15 years to rule out any cancer risk.

11.4. GENE THERAPY IN THE UTERUS

The above section seems to demonstrate that it is extremely difficult to insert a corrective gene into the cells of patients in such a way that it functions well and that the recipient of the gene gets better and stays healthy. According to Mels Sluyser (1999), a well-known cancer researcher as well as an artist of some distinction[75], the question is whether the gene

therapy would be more successful if it were carried out prenatally, on the embryo in the uterus. Embryonic cells grow faster than adult cells and so take up "foreign" DNA more easily. Moreover, the immune system in the embryo is not yet fully developed, so there will be less risk of the "foreign" substance being rejected. Aside from the fact that this form of gene therapy begs all kinds of ethical questions (for example, whether embryos should be manipulated), there are also extra risks associated with it. If the treatment is not performed well, the embryo and, potentially the mother, may develop an infection which could lead to a miscarriage. Another concern is that the inserted gene ends up not only in the intended tissues but also in other tissues of the unborn child, for example, in the bone marrow, where it could cause damage.

The most controversial issue in the discussions on gene therapy is the possible danger for future generations. Such criticism is levelled not only at the treatment of the unborn child, but also at gene therapy in general. There are clearly only risks for future generations if the administered gene also gets into the reproductive cells, i.e. into the ova of the woman or the sperm of the man. It does appear that DNA can go from one cell to another, so in principle a little of the "foreign" DNA could reach the reproductive cells. However, it has been calculated that the risk of this causing congenital disorders in the offspring is extremely slight. What is clear is that much more research is needed to define good treatment protocols. Only then will an informed decision be possible on the question of whether or not it is acceptable to conduct gene therapy on an unborn child.

[75] www.mels-sluyser.com/Nederlands/overmels.html

The debate on preconceived genetic modification of reproductive cells is a whole different ball game, particularly when it involves the insertion of desirable characteristics (intellect, musicality, etc.) or the 'erasing' of undesirable characteristics (baldness, red hair, etc.); that takes us into the arena of eugenics. 'Made-to-measure children' may well be a great topic for conversation today, but the reality, should society ever find that acceptable, is a long way off.

PRECONCEIVED GENETIC MODIFICATION

I WOULD LIKE TO ORDER ONE BABY TO GO …

… MEDIUM SIZE, LOW ON FAT WITH EVERYTHING ON IT!

The previously mentioned Emeritus Professor Piet Borst was very clear about this in his column *Eugenetische oprispingen* (Eugenic burps) of 21 October 2000. "I don't think that anyone able to read this column (eight years and older) will ever witness this, because all those carelessly proposed procedures are not feasible. There are two major theoretical problems in changing our reproductive cells (aside from all the technical obstacles): directed change of a number of genes in the same ovum/sperm is impossible, nor is there any theoretical solution for this problem in sight; altering the expression of complex characteristics like intellect requires a knowledge of genes and gene interactions that lies outside our imaginary powers." Although technological advances often move at a faster pace than we believe possible, we completely agree with this statement, never mind the discussion as to whether or not it is even desirable.

11.5. NOT EVERYTHING CAN BE TREATED (YET)

Even if the genes responsible for all congenital diseases were known, the use of gene therapy would still be limited. This was obvious right from the early days of gene therapy (Mariman, 1994). Some basic knowledge of human genetics is invaluable in order to be able to understand this, and you can find this in Textbox 11.3. In principle gene therapy can cure mainly the recessive genetic diseases. The dominant or recessive nature of an inherited characteristic is already determined at fertilisation. The sperm cell delivers a complete set of genes to the ovum, which has a complete set of its own. So a fertilised ovum and the new cells resulting from this, excluding the reproductive cells, contain a copy of each gene from the father and the mother. The symptoms of a recessive genetic disease only manifest if both copies of a gene are defective. As long as one copy

Chromosomes carry genes and regulate cell activity. They are made up of DNA with RNA and proteins. It is assumed that every chromosome has a double helix which consists of two strands of complementary DNA (see also Textbox 1.1). The number of chromosomes per cell nucleus determines the type. If there is one of every chromosome, it is called a haploid cell. If there are two of every chromosome (so-called homologous chromosomes), then it is a diploid cell. Humans are diploid and have two sets of 23 chromosomes, thus 46 in total, of which one pair is sex chromosomes. Cell division is the process whereby a cell splits into two daughter cells. During this process the chromosomes duplicate – whereby the two strands of DNA form each other's matrix - and then split up during a process called mitosis, so that each daughter cell gets a package of chromosomes identical to the parent cell. Meiosis or reductional division, in contrast, is the cell division process that leads to the formation of daughter cells with the complementary half (or halves) of the genetic material of the diploid parent cell (2n). The haploid cells thus formed are sex cells (gametes) which again produce cells via fertilisation (union of a male and female gamete) with a complete set of chromosomes (2n). These cells have characteristics of both parent cells whereby the relationship between them is determined by a simple dominant/recessive relationship (Mendel's Laws) between the alleles.

An allele is one of the two possible forms of a gene in a diploid cell. The alleles of a specific gene occupy the same place (locus) on homologous chromosomes. A gene can assume different forms (alleles) due to mutations. A gene is homozygotic if the two loci have the same alleles and heterozygotic if the alleles are different. If there are two different alleles present, one of them (the dominant allele) suppresses the effect of the other (the recessive allele). The allele that determines the normal form of the gene is usually dominant, while the mutated cell is usually recessive. Most mutations are therefore expressed in the phenotype (outer appearance) if it is homozygotic, i.e. if both alleles are mutated.

---

is healthy, the disease will not be present. So the insertion of a healthy copy in the cells of a patient with a recessive disease may lead to an improvement or a cure. In dominant inherited diseases, however, the symptoms always manifest when there is one defective gene copy, even if there is a healthy second copy of the gene in the cells. In this case, therefore, the addition of another healthy copy will not lead to a cure. Maybe in the future, depending on the nature of the defect in the dominant gene, other genetic techniques, for example gene silencing, may offer a solution for some of the dominant inherited diseases. In addition to recessive and dominant genetic disorders, where there are defective copies of a specific gene, there is a big group of complex genetic diseases caused by a combination of different poorly functioning proteins and adverse environmental influences. Well-known examples of this are rheumatic disorders and the frequently occurring skin disease psoriasis. Several genes at a time would have to be corrected in order for these sorts of diseases to be effectively treated with gene therapy. This is a particularly challenging and, for the time being, unfeasible exercise.

11.6. GENE DOPING

The term gene doping is relatively new. It first surfaced in the media towards the end of the 1990s, when the first publications about muscle-strengthening genetic modification of rodents began to appear. If you "Google" the term *gene doping* now, you will get thousands of hits.

Right from the first gene therapy experiments in the early 1990s, researchers were discussing the possibility of strengthening muscles by inserting new genes and experiments were already being carried out on patients with a congenital muscular disorder. The first results, however, were disappointing. Injections of myoblasts (the progenitors of muscle cells), to which an intact gene had been added outside the patient's body, did not strengthen the muscles of young people with Duchenne muscular dystrophy. The first genuinely positive results of muscle-strengthening gene therapy were seen in the late 1990s with mice and these led to discussions about gene doping. Journalists call such transgenic mice, half affectionately and half warily, *Schwarzenegger mice*. A more recent, spectacular example of what gene doping can do to mice was published in the online journal *PLoS* in August 2007 by Se-Jin Lee of the Johns Hopkins University in Baltimore. He describes a genetic procedure aimed at influencing the formation of two proteins which regulate muscle

---

TEXTBOX 11.4.

The German muscleman.

In June 2004 the New England Journal of Medicine reported the existence of a German muscleman (Van Caulil, 2004). When he was seven months old he could already stand and as a mere four-year old he could hold dumb-bells weighing three kilos in outstretched arms. So the 2020 Olympic Gold for weightlifting is virtually a given for Germany. But this is not the result of the notorious East German doping programme of that time. It is nothing more than Mendelian inheritance. His mother was a strong athlete, with one version of a defective myostatin gene. Her precious son got

mutations in both versions of the gene. The father is not known to the doctors.

THE GERMAN MUSCLEMAN

growth, namely myostatin and follistatin. Myostatin inhibits muscle growth, while follistatin deactivates the effect of myostatin by binding to it. The aim of the procedure was to deactivate the myostatin gene and continually activate the follistatin gene. The resulting transgenic mice acquired four times as much muscle mass. This result was a lot better than expected and led to the suspicion that a few other unknown mechanisms play a role in muscle formation. Lee advises athletes not to experiment with this, but he hopes to be able to help people suffering from muscular dystrophy, AIDS or cancer.

The tremendous impact on muscle growth of deactivating the myostatin gene was nothing new. The Belgian Blue cattle are blatant proof of how effective it can be. These animals are popular among cattle breeders: they have more muscle mass, a stronger skeleton and less fat. All without any training - they are lethargic beasts! The myostatin gene of these cattle contains an error, which means that they can't make myostatin. Moreover, the cows have traditional breeding to thank for this phenomenon, not gene therapy or genetic modification. A myostatin deficiency can also turn a man into Samson (Textbox 11.4).

The transgenic musclemen are not only arousing interest in the meat industry (Textbox 11.5). The WADA, the world authority in anti-doping, is also eager to know more. As far as is known gene doping has not yet been used in the sporting world. And yet it has been on WADA's official black list since 1 January 2003. That's the reverse way of doing things compared with most other forms of doping. WADA had a broad definition for gene doping (Schjerling, 2004): "the non-therapeutic use of cells, genes, genetic elements, or of the modulation of gene expression, having the capacity to improve athletic performance." Meanwhile it has been revised to: "The transfer of cells or genetic elements or the use of cells, genetic elements or pharmacological agents to modulating expression of endogenous genes having the capacity to enhance athletic performance, is prohibited." Schjerling is a researcher at the Muscle Research Centre in Copenhagen, an institute that examines the question of how sportsmen and women could raise their level to achieve medal status with gene therapy. In November 2003 the Rathenau Institute organised a symposium on the 'makeable' man. This institute shows the impact of science and technology on our daily life and charts the dynamics of this impact by conducting independent research and debate.[76] Schjerling said at this symposium: "I don't expect to see any genetic doping at the 2004 Olympic Games in Athens. ... The risks at this stage are too great. ... Research is not advanced enough." During the Olympic Games in Athens there were frequent statements in the media expressing the expectation that this would be the last Games without gene doping. Hidde Haisma, a professor in Therapeutic Gene Modulation at the University of Groningen, also supported this conclusion in 2004, having spent that year analysing the possibilities and risks of gene doping at the request of the Netherlands Centre for Doping Issues (NeCeDo). However, it now seems that the 2008 Olympic Games

[76] www.rathenau.nl

TEXTBOX 11.5.
Hormone mafia becomes gene mafia.

For years Willem Koert has been a scientific journalist on the weekly paper at Wageningen University. In the issue that appeared on 14 October 2004 he wrote an article on gene doping with the same headline as this textbox. The subtitle of the article was as follows: "Illegal fiddling with genetic modification untraceable." The summary stated: "Doping detectives think that the first gene technologically changed top athletes will be making their appearance within the next few years. Will the hormone mafia follow this example and feed cattle using 'gene doping'? The clandestine potential of an experimental technology."

At the moment the hormone mafia are still using old-fashioned preparations such as anabolic steroids and clenbuterol, but who will be able to stop them from using gene technology now that the possibility has come enticingly close? Researchers from John Hopkins University have published prolifically about the myostatin gene. Due to a defect in that gene "double-muscle cows", such as the Belgian Blues, make none or hardly any of the hormone protein myostatin and as a result acquire extraordinarily big muscles. Using gene technology, this gene could in principle be deactivated in healthy cows, but there remains considerable doubt as to whether that alone will increase the muscle growth and/or meat production in these so-called 'knock outs'. There is already a lot of practical experience with animals who by nature have one myostatin gene that doesn't work or works inefficiently. For example, it is well known that there are varieties with a poorly functioning myostatin gene that do not suffer any health problems. And yet breeding programmes are thwarted by dystocia. This means that the calves become so big in the uterus that it is impossible to bring them into the world naturally. A Caesarean section is always required. Randomly intervening in the genome of animals in order to deactivate the myostatin gene is therefore not a real option, never mind all kinds of other possible complications. So we are in complete agreement with Koert. Consequently, to end with, his central magnified, fairly crass statement: "If idiots start experimenting with this technology in underground labs, I dread to think what will happen." And so do we!

in Beijing were also free of gene doping. Of course, we can't be absolutely certain because there are no tests yet. But these are being developed with some degree of urgency. WADA finances most of these research projects. Françoise Lasne, who as we saw in Chapter 10 has been responsible for so much groundbreaking doping test research, has already shown that doping with the gene for EPO, injected into key muscles before a sporting performance, can be traced. She conducted the experiment on macaques. However, the complexity of the matter leads to the unfortunate suspicion that this is yet another case of lagging behind events.

11.7. GENE THERAPY: NOT YET A PANACEA OR A REVOLUTION

As mentioned earlier, more than 4,000 diseases and abnormalities, ranging from SCID to cystic fibrosis, are caused by a congenital defect in a single gene. Many other disorders, including cancer, heart defects, HIV and senility, are to some degree the result of postnatal damage to one or more genes. It is also clear that gene therapy is still far from being a panacea for all diseases caused by a defect in one single gene. Yet, in this chapter, we have also shown that there have been a few successes, albeit with eventual setbacks, in the treatment of some serious genetic abnormalities. For the time being this will not be the case for many other genetic disorders because there is limited or no knowledge about their genetic foundation. However, this situation is likely to change rapidly in future decades, now that the human genome has been essentially entirely mapped (see Chapter 13). The biggest task is yet to be tackled, though, and that is the identification of all genes, their functions and the way in which they operate. Only then will it be possible to point to the genes responsible for specific diseases. The technique itself will also have to be substantially improved, in order to properly insert corrective genes into the cell, so that they have a lasting desirable effect. In short, neither panacea nor revolution just yet - but hopefully, as Winston Churchill once said, it is the end of the beginning!

That success in gene therapy is still a long way off is evidenced by the fact that there have still been no recorded cases of gene doping. Sporting history is full of examples of athletes suffering premature death, cardiac dilation and other ailments in return for a higher chance of victory. A who-dares-wins approach! That was also the message from Mark Frankel of the American Association for the Advancement of Science (AAAS) at the gene doping symposium in St. Petersburg in June 2008 (Köhler, 2008). There he demonstrated that the huge financial stakes would guarantee that gene doping will be used as soon as possible. As Frankel says, sport, medicine and science, it's all business. And right in the middle, there's gene doping! We're very much afraid he may be right.

11.8. SOURCES

Bainbridge, J., Smith, A., Barker, S., Robbie, S., Henderson, R., Balaggan, K., et al. (2008). Effect of gene therapy on visual function in Leber's congenital amaurosis. New England Journal of Medicine, 358(21), 2231-2239.

Blaese, R., Culver, K., Miller, A., Carter, C., Fleisher, T., Clerici, M., et al. (1995). T lymphocyte-directed gene therapy for ADA-SCID: initial trial results after 4 years. Science, 270(5235), 475-480.

Cavazzana-Calvo, M., Hacein-Bey, S., Basile, G., Gross, F., Yvon, E., Nusbaum, P., et al. (2000). Gene therapy of human severe combined immunodeficiency (SCID)-X1 disease. Science, 288(5466), 669-672.

Couzin, J., & Kaiser, J. (2005). Gene therapy - As Gelsinger case ends, gene therapy suffers another blow. Science, 307(5712), 1028-1028.

Edelstein, M. L., Abedi, M. R., & Wixon, J. (2007). Gene therapy clinical trials worldwide to 2007 - an update. Journal of Gene Medicine, 9(10), 833-842.

Edelstein, M. L., Abedi, M. R., Wixon, J., & Edelstein, R. M. (2004). Gene therapy clinical trials worldwide 1989-2004 - an overview. Journal of Gene Medicine, 6(6), 597-602.

Friedmann, T., & Roblin, R. (1972). Gene therapy for human genetic disease? Science, 175(4025), 949-955.

Hacein-Bey-Abina, S., von Kalle, C., Schmidt, M., Le Deist, F., Wulffraat, N., McIntyre, E., et al. (2003). A serious adverse event after successful gene therapy for X-linked severe combined immunodeficiency. New England Journal of Medicine, 348(3), 255-256.

Köhler, W. (2000, 29 January). Fatale haast; meeste afweerreacties op gentherapie worden niet gemeld. NRC.

Köhler, W. (2008, 26 July). Epodope. NRC.

Mariman, E. (1994). Gentherapie gaat erfelijke ziektes te lijf. Chemisch Magazine (March), 104-106.

Peng, Z. (2005). Current status of gendicine in China: recombinant human Ad-p53 agent for treatment of cancers. Human Gene Therapy, 16(9), 1016-1027.

Raper, S., Chirmule, N., Lee, F., Wivel, N., Bagg, A., Gao, G., et al. (2003). Fatal systemic inflammatory response syndrome in a ornithine transcarbamylase deficient patient following adenoviral gene transfer. Molecular Genetics and Metabolism, 80(1-2), 148-158.

Rosenberg, S., Aebersold, P., Cornetta, K., Kasid, A., Morgan, R., Moen, R., et al. (1990). Gene transfer into humans--immunotherapy of patients with advanced melanoma, using tumor-infiltrating lymphocytes modified by retroviral gene transduction. New England Journal of Medicine, 323(9), 570-578.

Schjerling, P. (2004, 1 November). With good comes the bad: the reality of gene doping; the misuse of gene therapy by elite athletes is inevitable. Genetic Engineering News.

Sedlak, B. J. (2003, 15 May). Possibilities move forward for gene therapy; gene activity therapeutics enter phase II trials. Genetic Engineering News.

Sluyser, M. (1999, 17 April). Gentherapie in de baarmoeder. Telegraaf.

Van Caulil, G. (2004, 17 September). Transgene strijd – Gentherapie zal de weg volgen van menig ander geneesmiddel: een toepassing als doping. Bionieuws, pp. 8-9.

"People are entitled to disagree with xenotransplantation, but then they should register as organ donors."

The above statement was made by Guido Persijn, former Medical Director of the Eurotransplant Foundation, an international organisation that coordinates organ donation and transplantation. Xenotransplantation is the transplantation of organs, tissues or cells from one species of animal to another. This chapter will look at transplantations between animal and humans. Xenotransplantation is one possible solution for the organ donor shortage in the area of transplant medicine[77]. However, there is still a ban on this type of procedure because of the lack of clarity about the sort of risks entailed. The natural rejection responses to cross-species components still create insurmountable problems. The transfer of viral DNA with, as yet, unpredictable consequences is also another matter that requires due attention. The various facets of this topic will be discussed in this chapter, as will the question of whether or not xenotransplantation is ethically responsible. We'll begin with the history of xenotransplantation, which has its origins in a dark past.

XENOTRANSPLANTATION IS GREAT
STUFF FOR BACHELOR PARTIES

I THOUGHT A BUNNY SUIT
WOULD BE ENOUGH!

---

[77] www.eurotransplant.org/?id=xeno

12.1. THE HISTORY OF XENOTRANSPLANTATION: A SHOCKING PAST

Don't take your organs to heaven; heaven knows we need them here.

This quotation, whose origins are unknown to me (JT), hangs on the wall in my sister's bathroom. For more than ten years, she has been hosting one of my brother's kidneys (see Textbox 12.6). It was a perfect match! So you'll understand my particular interest in this topic. I wrote the first draft of the Dutch version of this chapter in 2001 during a sabbatical in Lausanne. It lay untouched until 2007 when in December of that year my co-author (YZ) updated it. On 28 October 2008 I began the final revision of the Dutch version by reading an article on the history of xenotransplantation, written by Deschamps *et al.* (Deschamps, Roux, Sai, & Gouin, 2005). Because of my personal interest in the subject of organ donation, the article read as a real horror story, so incredibly thrilling, that I felt almost guilty for reading it 'in the boss's time'. It became the most important source of information for this chapter. The Dutch book was published in November 2009. Although we did quite a bit of revision, it remains also one of the main sources for this English version.

Long before there was any insight into xenotransplantations of whatever kind, there were stories in folklore about creatures that were half-man, half-beast (chimeras). Pre-historic cave drawings seldom showed people, but in the Lascaux caves in France there is a drawing of a man with a bird's head (circa 15,000 BC). The first description of what can be defined as xenotransplantation comes from Indian mythology, in a Sanskrit text from the 12th century BC (Textbox 12.1).

TEXTBOX 12.1.

The first xenotransplantation.

Shiva and Parvati are two Gods from Indian mythology. According to the legend, their child Ganesha was born while Shiva was out hunting. As in so many myths, Ganesha was born a giant. When Shiva returned home and saw his wife with this gigantic 'stranger', he beheaded him. Parvati informed him that he had just killed his own son, and threatened to destroy the universe if Ganesha was not resurrected. Shiva, who wasn't able to re-attach Ganesha's head, ordered his servants to bring him the head of the first living creature they encountered. And so Ganesha was given new life with the head of an elephant.

Source: Shutterstock

Going back many centuries before Christ there were already descriptions of transplantation experiments with people. For example, Susrata, an Indian surgeon, was said to have used pieces of skin, about 600 years before the calendar era, to replace noses that had been cut off (often as a punishment). The most notorious example of early transplantations is probably 'the miracle of the black leg' from the 3rd century AD. In Rome two Syrian doctors, Cosmas and Damian, amputated the gangrenous leg of a verger. They attached in its place the leg of a dead black man. All these early attempts were transplantations from human to human (allotransplantation); all were probably not very successful, even though the species barrier had not been crossed. However, what they

clearly show is that the idea of extending or enhancing life, by replacing failing or missing organs with organs from a human donor, was already around a long time ago.

The first xenotransplantations were carried out at the beginning of the sixteenth century with cells and tissues from bone, skin, and testicles, etc., the latter playing a particularly important role (Textbox 12.2). Organs were only used much later on, because for a long time there was no technique to keep the bleeding under control once the diseased organ had been cut out, and no way to restore blood circulation after the transplant.

In 1668 the Dutchman Job van Meekeren reported a successful xenotransplantation with bone, performed by a Russian who used a piece of dog skull to repair a human skull. His claim that this had never been done before was later refuted. In 1501 the Iranian physician Muhammad Baha' al-Dawla published medical notes in *The Quintessence of Experience.* In it he described the surgical removal of a piece of skull that was infected and replaced by a piece of dog bone, the brain being protected by a piece of cucumber. He also reported that in Herat, Afghanistan, the Indian surgeon Ala-ul-Din had used fresh canine skin to replace all the skin on the head of a patient suffering from eczema. These early experiences were followed by many other similar primitive experiments. The first person to describe it as a transplantation was the Scottish surgeon John Hunter in 1778, when he wrote of a transplantation of a human tooth to the comb of a cockerel.

SOMETHING WENT WRONG

I'M SORRY, THERE WAS A SMALL MIX UP WITH YOUR ALLOTRANSPLANTATION.

YOU'RE SORRY?

Xenotransfusions are also centuries old. The transfusion of lamb's blood to a 15-year old boy on 15 June 1667 is the first documented account of such a procedure. This was performed by the French doctor Jean-Baptiste Denis, King Louis XIV's physician, and the surgeon Paul Emmerez in Paris. Less than half a year later, on 23 November, Richard Lower carried out the same experiment in London on the 22-year-old Arthur Coga, also successfully. Several other transplantations followed, but with less success, and on 10 January 1670 they were banned by the French parliament, quickly followed by the English parliament and then by the Pope. Despite this ban, documents describing xenotransfusions continued to appear, even until quite recently. In 2000 the Indian Dhani Ram Baruah administered more than quarter of a litre of pig's blood to the 22-year-old Hussan Ali, who was suffering from severe anaemia. Ali was discharged from hospital four weeks later. Tests confirmed that he had non-human blood cells in his bloodstream. A few years ago, fresh blood shortages also led to calls to reconsider xenotransfusions. Artificial oxygen carriers

---

TEXTBOX 12.2.
Human rejuvenation transplants.

In 1889 the French-American doctor and physiologist Charles-Edouard Brown-Séquard injected himself subcutaneously with an aqueous extract of crushed testicles from dogs and Guinea pigs. These injections were intended to restore his physical strength and capacities that were diminishing due to the ageing process. In so doing the 72-year-old Brown-Séquard invented opotherapy, a treatment using bodily fluids and a forerunner to endocrinology. Since then, numerous medicines based on crushed animal organs have come onto the market; extract of thyroid and pancreas are still available. Serge Voronoff turned this therapy into a surgical procedure. Born in Russia in 1866, Voronoff acquired French citizenship in 1895. He wanted to rejuvenate men by means of xenotransplantation with the testicles of chimpanzees and baboons. On 12 June 1920 he carried out the first procedure on a man, using the testicles of a chimpanzee; slithers of testicle were inserted into the scrotum. Three years later 43 men had undergone this operation and in 1930 that figure rose to 500. Women received an ovary from female apes for the treatment of menopause. Yet more shocking is the fact that he inserted a human ovary into a female chimpanzee (Nora) and inseminated her with human sperm; so much for ethics!? The insemination was unsuccessful, but Nora did get the leading role in the 1929 novel "Nora the she-monkey becomes a woman".

During his career Voronoff was concerned about the adequate supply of donors (monkeys), and considered setting up monkey farms in French Guyana in which to breed monkeys for export. Vilified by the scientific community and the public, Voronoff stopped performing after 2000 xenotransplantations. He died in 1951. It should, however, be noted that he was the first person to draw attention to the problem of donor shortages.

for blood transfusions have also been attracting attention in research circles in recent years.

The most important criterion for successful organ transplantation is the restoration of the vascular tissue of the organ in question by the stitching together of the arteries. More than a century ago, the Frenchman Mathieu Jaboulay and his student Alexis Carrel pioneered this technique, using mainly kidneys. Kidneys were preferred because it was easy to prove the success of the operation: you just had to wait until the patient urinated!

In January 1906, they attached the kidney of a pig, slaughtered three hours before, to the inside of a woman's elbow. In the next day and a half they collected one and a half litres of urine, but on the third day they were forced to remove the kidney because of thrombotic symptoms. Three months later they repeated the same operation with the same result in a different woman, this time with a goat's kidney. These transplantations are often regarded as the first real xenotransplantation experiments, even as the first organ transplantations. In 1909, 1913 and 1923 other researchers carried out a xenotransplantation with a kidney from a macaque, a Japanese monkey and a lamb, respectively, but the results weren't much better. There then followed a period of 40 years without further attempts.

The failure of these first experiments was the direct result of a lack of means to prevent rejection by the immune system. When these means came in the early 1960s, the interest in (xeno)transplantations was reawakened. In 1964 a 23-year-old woman received a kidney from a chimpanzee. When she died nine months later, an autopsy revealed the only cause of death to be an acute imbalance of electrolytes. Nine months surviving without rejection of the liver delivered proof that xenotransplantation was possible in principle. Yet the results remain unimpressive and this nine-month survival is still a record. In 1976 the Swiss Jean-François Borel discovered cyclosporin A, a drug which was also expected to be capable of suppressing rejection reactions in xenotransplantations.

On 26 October 1984 the American Leonard Bailey carried out the operation that would become the most famous xenotransplantation in history. It concerned the premature twelve-day old *Baby Fae* who had a heart defect. She was given a baboon heart. The conditions seemed optimal, including the fact that cyclosporin A was now available. In addition, of the six available baboons, the one chosen gave the weakest response with white blood cells and would probably cause the least rejection reactions as a result. After eleven stable days, however, the first rejection symptoms were observed and 20 days after the xenotransplantation Baby Fae died. Much of the hope vested in xenotransplantation died with her and a moratorium followed *de facto*. The new anti-rejection drug, Tacrolimus, which was brought onto the market in 1992, did little to change this situation. The arrival of transgenic pigs the same year, however, did seem to herald change and a new era.

12.2. THE TRANSGENIC 'SPARE-PART PIG'

"The creatures outside looked from pig to man,
and from man to pig,
and from pig to man again;
but already it was impossible to say which was which."

George Orwell: Animal Farm; 1945

Source: Identim

Anyone looking at animals to solve the shortage of transplant organs, should actually focus on those species that are evolutionarily most closely related, e.g. primates (mammal group that includes monkeys, apes and human beings). Apes are therefore the recommended organ donors, especially since it now appears that there is a less than 1% difference in genetic make-up between chimpanzee and man. That would seem to imply that there would be few rejection symptoms and only slight differences in the organ functions. But, after several kidney transplants in the 1960s from chimpanzee to human, the entire medical community has agreed that ape organs are not a real option. This has to do with anatomical differences, animal ethics and practical problems with breeding, but probably more to do with the growing understanding that non-human primates are a source of viruses that can be or can become very dangerous for humans.

Pigs are now the animal of choice. The anatomy of their internal organs shows major similarities to those of humans. They are also available in abundance and have a relatively short reproduction cycle with big litters. Years of experience also show that pigs are relatively easy to breed in germ-free conditions. But from an evolutionary point of view, pigs are a lot further away from humans. This results in all kinds of acute and chronic rejection symptoms. Genetic modification of the pig and immune suppression in the organ recipient should overcome that. But even in the absence of these obstacles, there still remains the question of whether the physiology of the pig is similar enough to that of humans. It is, as yet, a largely unexplored area of xenotransplantation: the comparison of physiological and biochemical characteristics of man and pig; the subtle differences in hormone regulation, mineral concentrations and blood pressure.

The size of hearts, kidneys, lungs, liver and blood vessels in pigs are a good match with those of humans, making pigs ideal organ donors. But without special precautionary measures, xenotransplantation is guaranteed to fail. The rejection reaction is so strong, that the animal organ can be irreparably damaged within a few minutes. This sort of hyperacute organ rejection also occurs if a patient gets a human organ from a donor with a different blood group. Blood groups are determined by the nature of the sugar chain of the red blood cells (Figure 12.1). However, these blood group determinants are not only found in red blood cells; they are also present in most organs, cells and tissues in the body. In pig organs, as well as in the pig cells that line the inside of the pig's arteries, there are blood group determinants, which are recognised by our immune system as foreign, and for which we already have an antibody. This may be because we are confronted early on in life with bacteria which have the same sugar chain on the cell surface (Prather, 2007).

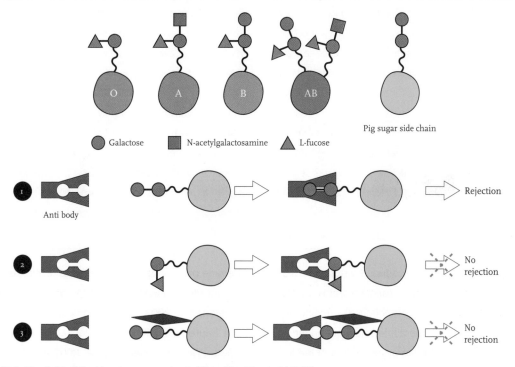

Figure 12.1. The fight of the blood groups, adapted from Van Zundert (1998).

organ from a donor with a different blood group. Blood groups are determined by the nature of the sugar chain of the red blood cells (Figure 12.1). However, these blood group determinants are not only found in red blood cells; they are also present in most organs, cells and tissues in the body. In pig organs, as well as in the pig cells that line the inside of the pig's arteries, there are blood group determinants, which are recognised by our immune system as foreign, and for which we

So our blood reacts directly to the pig organ, resulting in hyperacute rejection (Reaction 1).

Up until the turn of the century, getting around this immune reaction by modifying the blood group determinants in the pig was more of a theoretical than a practical or feasible solution. That's why, at the beginning of the 1990s, Imutran, a subsidiary of the pharmaceutical giant Novartis, tried a different approach. Researchers there succeeded in genetically modifying pigs so that they produced a human protein that could block the immune reaction between pigs' organs and human blood (Reaction 3). Testing with apes showed that the hyperacute rejection did indeed fail to occur. And yet, the pig organs failed in the long run. Preventing a hyperacute immune reaction was clearly not the only challenge. At a later stage the body throws in a whole army of antibodies and different processes into the fight to take out the 'intruders'.

Even when there is a "perfect match" between human donor and recipient (Textbox 12.6), a transplant only succeeds by at least temporarily suppressing the immune system with medication. However, it is still not known whether this medication is effective enough in the case of a xenotransplantation. Which is why there is lots of research taking place into these complicated immune processes, in order to better understand them so that more effective means to combat them can be developed. In 2007 recent progress, the state of affairs and future possibilities were looked at by Yang and Sykes (Yang & Sykes, 2007a, 2007b) in two leading journals. These data are not easy to summarise, but suffice it to say, there is still a long way to go. On the website[78] of Cytos Biotechnology we found a short and simple summary of the immune system (Textbox 12.3). Cytos Biotechnology is a Swiss company that is developing and commercialising a novel class of medicines – called Immunodrugs™. Immunodrugs™ are therapeutic vaccines intended for use in the treatment and prevention of common chronic diseases which afflict millions of people worldwide.

[78] www.cytos.com

So to start with, a successful xenotransplantation involves preventing hyperacute rejection (Reaction 1). The most obvious solution is to give the pig the sugar chain that corresponds to the determinant of our blood group 'O', because there are no antibodies for this sugar chain (Reaction 2). This means that only the extremity of the sugar chain, the galactose α-1,3-galactose epitope (an epitope is the location on the antigen which can be specifically recognised by antibodies that will bind to it) in the blood group determinant of the pig, will have to be changed: a fucose in place of a galactose. The enzyme that links the latter sugar unit (galactose), α-1,3-galactosyltransferase or α-1,3-GalT for short, must be deactivated and an extra enzyme must be 'incorporated' to ensure that the fucose attaches to the right location. Deactivating a gene entirely is always a complicated process, which until recently only had a reasonable chance of success in less complex animals such as mice, resulting in so-called "knock-out" mice. Since 1992, however, researchers have been working hard on the genetic modification of pigs for the purposes of xenotransplantation. In their review 'Xenotransplantation: The next generation of engineered animals' d'Apice and Cowan (2009) address the questions: What to remove? What to add? And how to do it? In their final section "Horses for courses?" they end with: "Sounds familiar ... so maybe a hurdler can run on the flat? Our approach is to try to build a multitalented pig and put him over a few different courses."

As mentioned previously, the English company Imutran (Cambridge) began research in the early 1990s on the genetic modification of pigs with a human gene. The gene in question codes for the protein hDAF (human decay-accelerating factor), one of the proteins that inhibits the so-called *complement activation* in humans (Siegert, Van Es, & Daha, 1996). Complement activation is the result of the cascade-like interaction of plasma proteins and membrane-bound proteins. There is a total of approximately 30 proteins in the complement system. The complement system traces intruders in the bloodstream and destroys them by drilling through their cell membranes. In a transplanted organ they bind to the endothelial cells, the 'inner lining' of the blood vessels. These cells are immediately destroyed. As a result coagulation factors are released from the underlying cell layers, causing the blood vessels to be blocked within minutes. Pierson III *et al.* (2009) have nicely pictured and described this process in their invited review article "Current status of xenotransplantation and prospects for clinical application" (Textbox 12.4). The complement system has a number of safety markers where the induced reaction can be stopped (this is necessary if a complement factor binds to the body's own structure); hDAF marks one of these points. This also applies to proteins coded as CD59 and MCP, as well as hDAF membrane-bound proteins[79].

On 23 December 1992 the first hDAF transgenic pig was born; it was named Astrid. Three years later the American company Nextran produced transgenic pigs that expressed two of this type of protein, hDAF

[79] www.ntvg.nl/node/290588/print

Coagulation is occurring continuously within the bloodstream, but is normally restrained by a network of inhibitory pathways involving endothelial proteins such as thrombomodulin and tissue factor pathway inhibitor (TFPI) (Panel A). Increased coagulation is normally initiated when endothelium retracts or becomes 'activated' by injury, in part because von Willebrand factor (vWF) is expressed and tissue factor (TF) is liberated into the circulation. The coagulation cascade then becomes amplified by the factors shown (VIIa/TF complex, IXa, and Xa) which in turn activate

thrombin. Thrombin amplifies the clotting cascade by (a) activating XIa (not shown), (b) activating platelets, (c) cleaving fibrinogen into fibrin monomers that form the primary clot matrix, and (d) activating factor XIIIa (not shown), which cross-links fibrin monomers into an insoluble clot. TFPI and thrombomodulin normally inhibit coagulation on healthy endothelium, while soluble antithrombins inhibit thrombin by forming a complex with its active site.

Porcine EC activation – whether by xenoantibodies, complement, or other factors – results in loss of natural anticoagulant proteins (TFPI, thrombomodulin) and acquisition of a procoagulant phenotype (Panel B). In addition functional incompatibilities in the coagulation system between pigs and humans cause

Human endothelium exposed to human blood

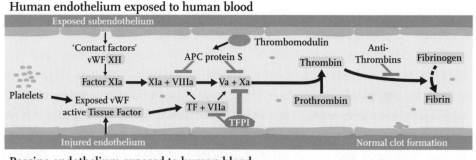

Porcine endothelium exposed to human blood

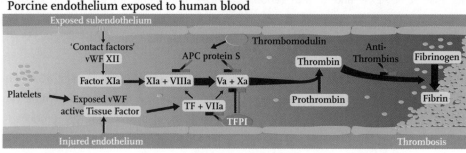

Figure 12.2. Dysregulated coagulation in pig-to-primate xenotransplantation, reproduced with permission (Pierson III et al., 2009).

both inappropriate or accelerated thrombin formation and inefficient restraint of clot activation. Our current hypothesis is that xenografts succumb to an otherwise insignificant humoral or cellular immune response which amplifies endothelial injury and intravascular thrombosis, and becomes manifest as thrombotic microangiopathy. The blue arrows designate cascade amplification steps, while the red lines identify loci of inhibition. The relative intensity of clot formation, the net product of coagulation pathway enzyme effects, is symbolised by arrow weight at the thrombin and fibrin steps. Pathways where pig endothelial proteins inefficiently dampen coagulation are indicated with hatchmarked red lines in Panel B. For simplicity, only the activated clotting factor intermediaries and key points at which regulation occurs are shown.

---

and CD59, with an even greater chance of stopping rejection. The Texan Robert Pennington owes his life to one of these pigs. In autumn 1997 the liver of this man, who was 17 at the time, suddenly failed. There was no donor available at the time. Dr. Marlon Levy, a transplant surgeon at Baylor University Medical Center in Dallas, offered to pump his blood outside his body through a transgenic pig's liver until a donor liver became available. One of the Nextran transgenic pigs, a sow later named *Sweetie Pie* by Robert, was transported to Dallas and slaughtered; its liver was connected outside Robert's body to his bloodstream. For nearly seven hours spread over three days until a donor liver was found, the pig's liver detoxified Robert's blood thereby saving his life[80].

The next major step forward was made in 2001, when two different groups announced that they had created transgenic α-1,3-GalT knock-out pigs and then cloned them. A year later, on 25 July 2002, the first four double knock-out pig clones (with both gene copies deactivated) were born at PPL Therapeutics, the company that created Dolly the Sheep. Since then there have been many more promising examples. For instance, a group at the University of Missouri-Columbia joined forces with Immerge Biotherapeutics to make an α-1,3-GalT knock-out of the Imutran pigs with an hDAF gene (Prather, 2007). This transgenic pig model was disseminated by the 'National Swine Resource and Research Center'[81]. Further modifications are still necessary before pigs' organs can be successfully transplanted to humans. The review of Klymiuk *et al.* (Klymiuk, Aigner, Brem, & Wolf, 2010) provides an overview of the transgenic approaches that have been used so far to generate donor pigs for xenotransplantation, as well as their biological effects in *in vitro* tests and in preclinical transplantation studies. As a future challenge they see the combination of the most important and efficient genetic modifications in multi-transgenic pigs for clinical xenotransplantation. Aigner *et al.* (Aigner, Klymiuk, & Wolf, 2010) review the selection of promoter sequences for reliable transgene expression for this purpose.

Xenotransplantation is still a very experimental

---

[80] www.pbs.org/wgbh/pages/frontline/shows/organfarm

[81] www.nsrrc.missouri.edu

procedure and no creature has yet survived a xenotransplanted pig organ for any length of time, not even a monkey. However, developments in recent years have been such that, particularly with transgenic pigs, pre-clinical studies with "pig-to-other-animal" xenotransplantations are running in various countries to further investigate the feasibility, and 'pig-to-human' xenotransplantations are starting to appear on the agenda again. Xenotransplantation of the insulin-producing islets of Langerhans from the pancreas is in particular expected to proceed quickly from pre-clinical to clinical phase (Schuurman, 2008). One reason for this is that less than five percent of the islets of Langerhans in pigs have the Gal epitope, meaning that the risk of hyperacute rejection is smaller than in other pig organs (Prather, 2007). Consequently, no α-1,3-GalT knock-out pigs are expected to be needed, in contrast to other pig organs. Rajotte (2008) states that, according to clinical research, the preparation and transplantation of the islets must be SAS - Safe, Affordable and Simple: (1) safe with regard to the transfer of pathogens (see following section), (2) affordable within our health-care parameters, and (3) simple and reproducible production of transplantable islets with a minimum of regulatory control.

Transgenic pig clones clearly signal the first breach of the rejection barriers. Yet there are still no clinical trials involving humans. There is still too little data available on the extent of the risk of transferring potentially very infectious viral DNA. Concerns about this are huge.

12.3. PANDEMIC RISKS

The last few decades have seen the spread of new infectious diseases such as Ebola, HIV, Creutzfeldt-Jakob, SARS and Mexican flu. These were probably animal diseases by origin, which have now become infectious for humans. This has raised great fears that xenotransplantation would exacerbate such mutations. Xenotransplantation is therefore regarded as a serious risk for public health, because it brings with it the risk of transferring swine pathogens, in particular viruses that are not endemic to humans (Louz, Bergmans, Loos, & Hoeben, 2008). If patients receive immune suppressants and still have no immunity, xenotransplantation can, in the worst-case scenario, lead to a global pandemic with a new life-threatening virus. Many exogenous viruses can be eliminated by pathogen-free breeding, by selection and vaccination of the donor animals and by adequate screening of the organs for xenotransplantation. However, O'Connell (2008) concludes that suitable facilities for looking after donor pigs still need to be designed. These will require lots of money for investment and maintenance and it will be a considerable time before donor pigs from a suitable facility are made viable for clinical use. In short, there's still a long way to go just with regard to facilities and protocols.

The biggest concern is about the pig endogenous retrovirus (PERV), of which there are several copies in the pig genome. This concern goes back to 1997 when it was shown that PERV could infect human cells if they were grown in a test-tube. Further indications were demonstrated in 2000 by means of experiments with mouse models. PERV was transmitted when islets of Langerhans were transplanted from a pig's pancreas to an NOD/SCID (non-obese diabetic/severe combined immune-deficient) mouse; NOD/SCID mice are mouse models that are diabetic and have a defective immune system. This proves that animal viruses can be potentially transferred during xenotransplantation to humans. Since then this has been a subject of concern and discussion, not only among health authorities. It has given rise to a precautionary approach, strict regulation and even a moratorium in many countries on clinical trials, at least with humans[82]. It is crucial that we remain aware of the fact that xenotransplantation combines possible advantages for the individual patient along with the risk of serious, large-scale, new infectious diseases (pandemics). Basically, PERV requires and is undergoing a thorough risk assessment at this time.

12.4. SOCIAL AND ETHICAL ASPECTS

In 1998 about two and a half million Dutch people registered their organs for transplant after their death. However, this generous gesture is a drop in the ocean. On the global level the supply of donor hearts, kidneys, livers and lungs has for years met only a fraction of the demand. In 2002, the number of people across the world registered as waiting for an organ was over 250,000. Less than a third of those waiting received a transplant[83].

And despite all the 'moped drivers without helmets' and major donor events, it doesn't look as though this situation will change any time soon. In Europe in 2005 only 3,540 kidney transplants were carried out, and the average waiting time was three years. The picture was no different in 2008. At that moment there were about 75,000 and 11,300 patients on the waiting list for kidneys in the US and Europe respectively (Sprangers, Waer, & Billiau, 2008). There have been no drastic changes since then and from what we know now, xenotransplantation is unlikely to change this scene for the time being.

The transplantation of animal organs to humans is ethically acceptable, according to a Dutch committee of the Health Council in its published opinion in January 1998 to the Minister of Public Health. But the committee also concluded that, before surgeons can routinely proceed to xenotransplants, problems with rejection and infection must first be resolved. The committee's opinion was largely consistent with those that had appeared earlier in the United Kingdom and the US. The British Nuffield Council of Bioethics stated that the breeding of pigs was the most acceptable solution for xenotransplantation. Breeding monkeys for this purpose was deemed unacceptable, mainly because of the greater risk of infection. Monkeys are more closely related to humans than pigs. Pathogens in monkeys can more easily adapt when they enter a human body, than bacteria and viruses from pigs. And yet there is still a risk of infection from pigs' organs, as we saw earlier in this chapter.

[82] www.fda.gov/BiologicsBloodVaccines/Xenotransplantation/default.htm
[83] www.eurotransplant.org/?id=xeno

In an interview in 2000, Maaike Werner, the press officer for the Dutch Society for the Protection of Animals until 2005, gave the following answer to the question of whether xenotransplantation is ethically responsible or not: "The Society for the Protection of Animals doesn't think xenotransplantation is ethically acceptable. We are worried about the emergence of a new bio-industry for organs. Just because the meat industry is accepted, doesn't mean that the organ industry should automatically be allowed. The meat industry has to meet certain criteria on animal welfare, the organ industry is very different. Just imagine a genetically modified pig living in a sterile room without daylight or straw before it has its throat slit. And many laboratory animals have already gone the same way. A pig's heart has a smaller capacity than a human heart, so they want to make a donor pig do conditioning exercises. Don't you think that's absurd?"

EXERCISES FOR PIGS...
ONE OF THE ABSURD ASPECTS
OF XENOTRANSPLANTATION?

Faced with the same question, the virologist Ab Osterhaus (Erasmus University Rotterdam) said: "I can't give a yes/no answer to that question. In terms of animal welfare, I think xenotransplantation should proceed under the right conditions. We breed pigs for meat already. But if I look at the risk of infection in the future, I'm not sure about it. It may be that we create new viruses using this technique; then I predict a doom scenario like AIDS. Xenotransplantation should only proceed very gradually under strict conditions. For example, you can start transplanting parts of organs and test them for viral infections. Either way, developments in this area cannot be stopped. So we'd better adopt the right approach and make sure that there's a plan in place should anything go wrong."

In an interview in the same year, Jan IJzermans, a transplant surgeon at the Erasmus University Medical Centre in Rotterdam, summarised the facts: "The problem is that the discussion is still vague, because there's no real idea of how great the dangers are. We have to weigh up the interests of the individual patient who could be saved with a donor organ, and the risk of a new epidemic. In contrast to many other controversial medical procedures, however, xenotransplantation cannot be dealt with simply as a personal choice, if the possibilities are there. Anyone with a pig's heart or kidney may constitute a risk for the entire health of the public."

At the turn of the last century, the Dutch Foundation for the Consumer and Biotechnology used a subsidy from the Ministry of Public Health, Welfare and Sport (VWS) to organise a public debate on xenotransplantation.

The aim of the debate was three-fold: to give the public information, let them form an opinion on the basis of this information, and then assess the opinion. On 10 November 2000 the bioethicist Egbert Schroten (University of Utrecht) and the then VWS minister Mrs Els Borst, opened this debate called *Should xenotransplantation be allowed?* It followed an information campaign on xenotransplantation that began in December 1999. During that time the press also devoted a lot of attention to this subject. In the first months of 2001 citizens were able to go and discuss xenotransplantation in various locations around the country. In addition, the science theatre Pandemonia staged the play "Dierbaar Leven" (*loosely translated this means 'life is precious', but there is a double meaning in the Dutch since "dier" means animal*) about xenotransplantation. The idea behind this was to get young people interested in the topic. There was a website (no longer available) allowing for a debate on the subject. The public debate was concluded at the end of April 2001 with a final meeting. In mid-2001 the final report was published by the foundation. The summary and conclusions are available online[84], unfortunately only in Dutch, but the gist of these is as follows. About half the people who interactively took part in the debate had no particular opinion for or against xenotransplantation. Proponents point mainly to the possibility of saving human lives and solving the problem of donor shortages. Opponents place more emphasis on the "makeability" of the body - the extent to which we are ready to interfere with the body. The

unnaturalness of the procedure, the crossing of human and animal, and possible adverse consequences for quality of life, were also important considerations. Compared with other forms of organ replacement (existing and experimental), xenotransplantation was found to be the least acceptable.

Many believed that xenotransplantation should only be allowed if there were no other ways of solving the donor shortage problem. The risk of infection was the most important concern, but for the majority of people this was not a defining enough reason to reject xenotransplantation. In any case, people didn't expect xenotransplantation to be used as long as there were still uncertainties about the risks. Opponents objected to the consequences for animals with regard to animal welfare, genetic modification and the fact that xenotransplantation is a new use of animals. Others found these consequences problematic too, but for them the purpose of the use (saving human lives) was more important. Nor did the latter group see any basic difference between this and using animals for other purposes, such as eating them.

Young people were much more concerned about surviving than adults. While adults were more inclined to accept the end of their lives, young people were more willing to calculate in certain downsides of xenotransplantation. In conclusion, we would like to mention that people also indicated the desire to be able to vote in the future as to whether or not xenotransplantation goes ahead. Crucial in this regard was clarity on the criteria used by the government and doctors when deciding on who gets which treatment.

[84] www.weten.nl/webzine/nummer4_2001/pdf/bruikbaar.pdf

That, then, was the final report on the public debate. A Dutch website that is constantly concerned with this theme is that of the working group on transplantation questions, which critically analyses all the medical, moral and risk aspects of transplantation[85]. There is also a lot of relevant information to be found abroad, for example, from Canadian supporters[86] and American[87] and British[88] opponents.

Little has happened in the Netherlands since the big debate. Xenotransplantation has received very little attention and not much has changed as regards the thorny points. The global picture is very similar. What we do see is that there are still many scientific publications and review articles appearing. George (2006) concludes in his article that xenotransplantation will probably remain controversial because of the complex nature of the medical, ethical and legal questions. If the scientific problems were to be solved, the decision to proceed with the clinical application of this technique would depend, in his view, on a collective decision based on ethical, regulatory and legal frameworks arising from a consensus. What is clear is that there is still much opposition. Googling Frankenstein in combination with xenotransplantation yields thousands of hits, generally not very friendly with respect to this topic.

In recent reviews, Professor Mariachiara Tallacchini (Tallacchini, 2008; Tallacchini & Beloucif, 2009) defines what she sees as a suitable ethical, social and regulatory framework for clinical xenotransplantation. She does so on the basis of an analysis of recent literature on regulatory questions concerning xenotransplantation. She concludes that the global scale on which the research is currently taking place, requires that some aspects of xenotransplantation should be reconsidered. Inadequacies and weaknesses in national legislation can, in her opinion, not only have undesirable local effects, but international implications as well. Conversely, the lack of international implementation of rules or "loose" interpretation of standards has a negative impact on groups and populations who are already disadvantaged, and may result in potential risks worldwide. Although specific subjects such as animal welfare and rights continue to be discussed, the most important aspect of the regulatory questions concerning xenotransplantation is increasingly shifting to the multi-faceted aspects of locality and globality, where space, as well as formal legislation, is created for non-governmental networks as potentially flexible and normative instruments. In short, if we the authors understand it correctly, there's still a long way to go.

So far, the religious angle has received little attention in the press and scientific literature. We have only been able to find one recent article on this (Bruzzone, 2008), which basically concludes that no religion has an official ban on xenotransplantation!

[85] www.stelling.nl/xeno
[86] www.islet.org
[87] www.crt-online.org
[88] www.uncaged.co.uk

12.5. IN CONCLUSION

Xenotransplantation clearly stirs society's conscience. This social anxiety is mainly to do with the risk that transplanting an animal organ to a human body may wake a sleeping virus. The pig genome may contain viruses that are harmless to the pig. All they do is sit in the DNA and replicate. Nothing more. It could be, with the emphasis on 'could', because nobody knows for sure, that these viruses mutate in a human environment into a variant that is harmful to humans. It would be unlucky indeed to have created with our own hands a brand-new viral infection to rival, for example, HIV.

THERE ALSO ARE SOME UNEXPECTED COMPLAINTS AGAINST XENOTRANSPLANTATION

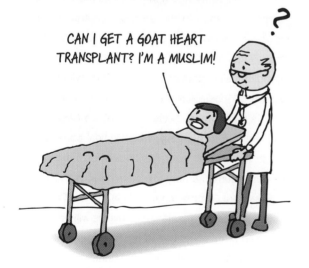

CAN I GET A GOAT HEART TRANSPLANT? I'M A MUSLIM!

No government, minister, doctor or patient would care to have that on their conscience. On the other hand, how can you gain more insight into the risks surrounding xenotransplantation if you completely ban research in this area? And even with research, there is still the question of how science can acquire real insight into the risks without exposure to those risks. It's not always possible to rationalise. It is fortunate that the decision rests not only on scientists' shoulders, but also on those of the whole of society.

Companies are also wrestling with this problem. Geron Bio-Med in Roslin, Scotland, a company trying in various ways to apply the cloning technique used to "make" Dolly the Sheep and to make money out of it, reported in mid-2000 that it was getting rid of its potential donor pigs. The company was planning to focus on embryonic stem cells that can be converted in the laboratory into specialised human cells and tissues for all sorts of medical applications (Chapter 14). The hope is that whole organs can be made in this way. Human, not animal, organs - from our own stem cells. But that's still in the future. In an interview in 2005, leading stem cell researcher Christine Mummery said that she wouldn't bet all her money on stem cell research being successful in this area. "If you need whole organs, xenotransplantation will be the only option for a long time to come", she predicted.

Yang and Sykes (Yang & Sykes, 2007b) say in their review article that considerable money and effort is being invested in alternatives to xenotransplantation. Artificial organs (Textbox 12.5) and mechanical devices may offer a potential solution for some organ failures, but in the near future they don't have the potential to supplant transplantation as a long-term curative therapy. Likewise organ and tissue regeneration on the basis of stem cells is very promising, but according to Yang and Sykes a good many years of research will be needed before the clinical phase can start. They therefore conclude that xenotransplantation may well be the current solution for the lack of donors.

This chapter has demonstrated that xenotransplantation is still a tricky business, both technically and ethically. The technical problems around rejection can probably be overcome in the relatively short term, but we certainly haven't read or heard the last word on the ethics and the risk of new viruses. Only when the most serious rejection hurdles and viral risks have been eliminated, can the practice really show whether complete pig organs can fulfil their replacement function as desired, or whether they will simply act as a short-term transition for the transplantation of human organs. This will obviously vary according to the organ in question: the heart has a less complex biochemical interaction with the body than the kidneys or liver. But this doesn't detract from the fact that more research is definitely needed in the area of xenotransplantation, however much the opinions of opponents and proponents differ. Maybe other options, such as organ breeding using stem cells, will catch up with the "spare pig parts" possibility before it comes into practice. This is surely more likely now that we have come to the end of the Bush era, when ethical considerations got in the way of federal financing for most research involving human embryonic stem cells. With the arrival of President Obama, change is in the air. Than again, maybe the xenotransplantation cynics will be right after all: "Xenotransplantation is the future and always will be." Only time will tell if they are right.

[89] www.willemkolffstichting.nl/index.php?phm=1

As I (JT) mentioned previously, in mid-1997 my sister suffered acute renal failure in both kidneys when she was just 50. This was followed by a long and miserable period, in which her close family repeatedly feared for her life. After more than a year, her condition stabilised and she had to undergo dialysis five times a day at equal intervals. Apart from all the other complaints, this was hardly an appealing state to be in. So the thought of a transplant wasn't far from our thoughts. In the first instance, the specialists thoroughly investigated the possibility of using a family member as a donor. My brother seemed to be the best "match" for my sister; in fact, the perfect match. Without thinking twice he donated one of his kidneys to my sister. By the beginning of 1999 it was a 'fait accompli'. The transplant had proceeded successfully and for more than ten years now my sister has been able to lead a relatively normal life with minimum use of medication; all because of a perfect match! It all makes you think differently about the alternatives. Not everyone who's on a donor waiting list is fortunate enough to have such a brother!

12.6. SOURCES

Aigner, B., Klymiuk, N., & Wolf, E. (2010). Transgenic pigs for xenotransplantation: selection of promoter sequences for reliable transgene expression. *Current Opinion in Organ Transplantation, 15*(2), 201-206.

Bruzzone, P. (2008). Religious aspects of organ transplantation. *Transplantation Proceedings, 40*(4), 1064-1067.

d'Apice, A., & Cowan, P. (2009). Xenotransplantation: The next generation of engineered animals. *Transplant Immunology, 21*, 111-115.

Deschamps, J. Y., Roux, F. A., Sai, P., & Gouin, E. (2005). History of xenotransplantation. *Xenotransplantation, 12*(2), 91-109.

George, J. F. (2006). Xenotransplantation: an ethical dilemma. *Current Opinion in Cardiology, 21*(2), 138-141.

Klymiuk, N., Aigner, B., Brem, G., & Wolf, E. (2010). Genetic modification of pigs as organ donors for xenotransplantation. *Molecular Reproduction and Development, 77*(3), 209-221.

Louz, D., Bergmans, H., Loos, B., & Hoeben, R. (2008). Reappraisal of biosafety risks posed by PERVs in xenotransplantation. *Reviews in Medical Virology, 18*(1), 53-65.

O'Connell, P. (2008). The rationale and practical issues for the maintenance of clean herds for clinical islet xenotransplantation. *Xenotransplantation, 15*(2), 91-92.

Pierson III, R., Dorling, A., Ayares, D., Rees, M., Seebach, J., Fishman, J., *et al.* (2009). Current status of xenotransplantation and prospects for clinical application. *Xenotransplantation, 16*(5), 263-280.

Prather, R. S. (2007). Targeted genetic modification: Xenotransplantation and beyond. *Cloning and Stem Cells, 9*(1), 17-20.

Rajotte, R. V. (2008). Moving towards clinical application. *Xenotransplantation, 15*(2), 113-115.

Schuurman, H. (2008). Regulatory aspects of pig-to-human islet transplantation. *Xenotransplantation, 15*(2), 116-120.

Siegert, C., Van Es, L., & Daha, M. (1996). Het complementsysteem en de klinische gevolgen van stoornissen. *Nederlands Tijdschrift voor Geneeskunde, 140*, 2268-2273.

Sprangers, B., Waer, M., & Billiau, A. (2008). Xenotransplantation: where are we in 2008? *Kidney International, 74*(1), 14-21.

Tallacchini, M. (2008). Defining an appropriate ethical, social and regulatory framework for clinical xenotransplantation. *Current Opinion in Organ Transplantation, 13*(2), 159-164.

Tallacchini, M., & Beloucif, S. (2009). Regulatory issues in xenotransplantation: recent developments. *Current Opinion in Organ Transplantation, 14*(2), 180-185.

Van Zundert, M. (1998). Varkentje vol reserveonderdelen. *Chemisch Magazine,* pp. 298-299.

Yang, Y. G., & Sykes, M. (2007a). Tolerance in xenotransplantation. *Current Opinion in Organ Transplantation, 12*(2), 169-175.

Yang, Y. G., & Sykes, M. (2007b). Xenotransplantation: current status and a perspective on the future. *Nature Reviews Immunology, 7*(7), 519-531.

THE HUMAN GENOME PROJECT

"The probability of life originating from accident is comparable to the probability of the Unabridged Dictionary resulting from an explosion in a print shop."

Albert Einstein[90]

Charles Robert Darwin, author of *On the Origin of Species*, was born on 12 February 1809, which was why so much attention was paid to the creator of the evolution theory in 2009, and why that year was designated the International Year of Darwin. It was also a good excuse for supporters of Darwin, for creationists (people who believe in the biblical version of the origin of the world), and other religious believers, to revive the age-old discussion on the question of whether it is possible as a scientist to believe in God and the Bible, the "book of life". As the quote above suggests, even a great scientist like Albert Einstein wasn't an unconditional proponent of evolution theory, of the theory that everything just happened by blind chance. The sublime example of 'the natural order and precision' is the DNA, the genetic material in the nucleus of living cells.

LIFE ORIGINATING FROM ACCIDENT IS AS LIKELY AS CREATING A DICTIONARY FROM AN EXPLOSION IN A PRINT SHOP ...

... RESEARCHERS TAKE PROBABILITY CALCULATIONS EVEN FURTHER!

NATIONAL LIBRARY

[90] answers.yahoo.com/question/index?qid=20071020041348AAmsi18

13.1. THE HUMAN GENOME

The collection of DNA in a cell nucleus, which is contained in 23 pairs of chromosomes in the case of humans, is called the genome, and is regarded by some as the blueprint of life, the language of God, or in other words: "the other book of life"! The study of genomes is called *genomics*. The year 2001 is recognised as the year in which researchers succeeded in sequencing all the building blocks in the human genome (Section 13.3). It consists of two sets of three billion building blocks and there are four different building blocks, as we saw in the second chapter in Textbox 2.1. Only two percent of the genome is used for our 23,000 genes, which contain the code for constructing our proteins, the building blocks of life. The surprises (see Section 13.6) that this unravelling revealed, teach us primarily that the wise old philosopher Socrates was right all those centuries ago when he said, "The more you know, the more you realise you know nothing."

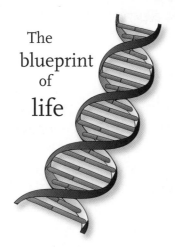

The blueprint of life

12.2 'THE BOOK OF LIFE'

At the very end of the 20th century the world witnessed an exciting race between the Human Genome Project - an international consortium of scientists from the public sector - and the commercial company Celera Genomics headed by Craig Venter. Who would be the first to decipher the human genome? On 26 June 2000 both parties jointly revealed, with much celebration and in the presence of the former American president Bill Clinton and the former British Prime Minister Tony Blair, that they had unravelled the genetic code - five years earlier than expected. Prematurely, in a sense, since neither had done much more than produce a rather rough draft. On 13 February 2001 there was yet more ceremony: both draft versions of "the other book of life" were published. *Nature,* the leading British scientific journal, devoted one hundred and fifty pages to it, including annotations. The article had 273 authors and contained the results of the Human Genome Project, directed at that time by Francis Collins. The American equivalent *Science* used up more than one hundred pages to print the data delivered by Venter's company, which began later than the public project, but caught up within a year. The two most prominent genome hunters, the deeply religious Christian believer Collins and the aggressive entrepreneurial scientist Venter, shared the honour. More than five years later both published a book in which they examined the human genome from their own personal perspective.

In *The Language of God: a scientist presents evidence*

for belief, Collins (2006) explains why it is that, in his eyes, science and faith should be able to walk the same path without conflict. He believes that creationists are denying truths that science has demonstrated. The essence of his argument is that the *Big Bang* and Darwinian evolution theory are sufficient to explain the creation of our world, including nature, the human body and the world of modern molecular biology in all its complexity, albeit with one exception: our perceptions of right and wrong. These perceptions can only be explained, according to Collins, by accepting that there was a second moment of creation. He believes that the language of creation can be read in the genome and in mathematics, two products of two moments of creation. The book is not an autobiography, but does contain personal anecdotes.

In contrast, Venter's book, *A Life Decoded. My genome: My Life,* is a full-length historiography (Venter, 2007). On the one hand he discusses in layman's terms what sort of information can be retrieved from genomes. On the other, he speaks about the actual information that was found in his own genome and the (possible) implications of that for his life. Each book is a justification of the path the authors chose, i.e. the path of a believer (scientist) and that of a (scientific) entrepreneur. Two paths in which the human genome did and still does play a central role.

An in-depth analysis of these two (semi) autobiographies and of a third one, i.e. *The common thread: science, politics, ethics and the human genome* by John Sulston with science writer Georgina Ferry as co-author (Sulston & Ferry, 2002), was published in 2008 by Hub Zwart (2008) (*Understanding the Human Genome Project: a biographical approach*). The two questions addressed:

1. What may we learn from these autobiographical sources about the dynamics of scientific change?
2. What is their added value in understanding science in general and the Human Genome Project in particular?

For non-philosophers the answers are not easy to understand and to summarise to a length fitting this chapter and we refer therefore to the original paper. What we do understand, though, is the answer of Francis Collins to the question at a press conference in San Francisco, February 2001, whether the sequencing warranted the Nobel Prize. He replied that it would have to be given to 3492 people to properly recognise everyone who had significantly contributed to this common effort. Zwart: "Although somewhat rhetorical, no doubt, autobiographical documents reveal that there is a kernel of truth in this reply."

In 2004 the Human Genome Project was officially wound down, despite the fact that some parts still hadn't been mapped. In subsequent years a lot of gaps have been filled and work is still being done on perfecting it. The American Department of Energy (DOE) maintains the website[91] which gives a detailed history of the project and the milestones, objectives and results. The question remains, what do we really know and understand about the human genome, and what can we do with it, particularly in the area of health care.

[91] www.doegenomes.org

DNA analysis

1. In order to determine the DNA sequence, it is first cut into fragments

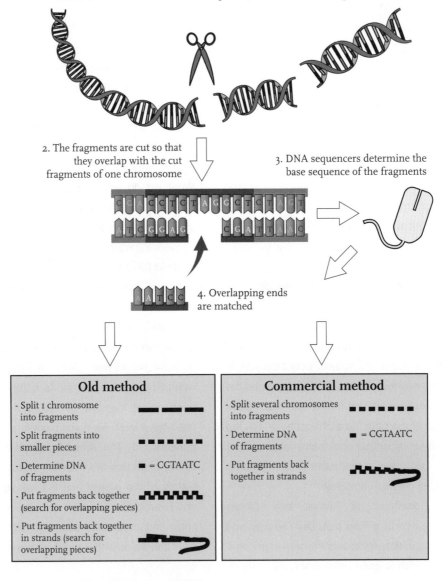

2. The fragments are cut so that they overlap with the cut fragments of one chromosome

3. DNA sequencers determine the base sequence of the fragments

4. Overlapping ends are matched

Old method
- Split 1 chromosome into fragments
- Split fragments into smaller pieces
- Determine DNA of fragments   ■ = CGTAATC
- Put fragments back together (search for overlapping pieces)
- Put fragments back together in strands (search for overlapping pieces)

Commercial method
- Split several chromosomes into fragments
- Determine DNA of fragments   ■ = CGTAATC
- Put fragments back together in strands

Figure 13.1. DNA analysis, adapted from Van der Laan (2000).

13.3. HUMAN GENOME SEQUENCING

The idea of determining the sequence of the bases (A, C, T, G) in human DNA emerged in the mid-1980s in the US and on 1 October 1990 the Human Genome Project (HGP) officially started. Work in this area was initially limited to the US, funded mainly by the government and a large national health-care organisation, but Europe, in particular the UK, and Canada and Japan later added their efforts. The mission[92] of the HGP was:

"To identify the full set of genetic instructions contained in our cells and to read the complete text written in the language of the hereditary chemical DNA."

In order to clarify the sequence of the approximately three billion human base pairs, the DNA was first cut into fragments. Two methods were used for this purpose (Figure 13.1). The slow, painstaking ('old') method which the researchers of the Human Genome Project opted for, and a "quick and dirty" approach employed later by Craig Venter. In the first method one chromosome was cut into large fragments, each of which was then individually cut into smaller fragments. The base sequence in these small fragments was determined automatically using a DNA sequencer. By cleaving with specific enzymes that very selectively break up the DNA chains in specific places, with a bit of playing around the original sequence can be found. The geneticist Craig Venter believed that it could all

be done much faster and much cheaper, and with that in mind he set up the company Celera Genomics in the late 1990s. Instead of using enzymes to cleave one chromosome into fragments, he used the shotgun approach to blast several chromosomes at a time into fragments of a few thousand base pairs. That can be done by vibrating the DNA to fragments with ultrasound or by forcing it under high pressure through a tiny opening. Using a large number of DNA sequencers these fragments are analysed, after which superfast computers fit all the pieces of the puzzle together into the original DNA chains, gratefully making use of the knowledge from the Human Genome Project, which is freely available on the Internet. However, though much faster, there is also the problem that the margin of error is much bigger. For this and other reasons, the two competing factions decided to collaborate at the beginning of 2000. By the end of June 2000 they thus jointly presented the first version of the human genome.

13.4. A NEW PARADIGM IN HEALTH CARE

Just before the turn of the last century, there was not only a final sprint to map the human genome, but also a great many Jules Verne-like predictions as to the effect modern biotechnology would have on health care in the 21st century. Such futuristic forecasts are not unusual with the advent of a new millennium. More than a hundred years earlier, in 1895, Lord Kelvin (an Irish-Scottish physicist, regarded as one

[92] www.accessexcellence.org/RC/AB/IE/Intro_The_Human_Genome.php

of the most important physicists of the 19th century) declared that flying machines were impossible. Eight years later the Wright brothers proved him wrong! In the same year, H.G. Wells (1866-1946) wrote 'The Time Machine', in which the main character travelled to the year 802,701 to see a world which boasts ideal preventive health care. Based on their timeline (Figure 13.2), Shamel and Udis-Kessler concluded in the December 1999 issue of Genetic Engineering News, that Wells was probably 800,600 years wrong! Certain that they could not make a bigger mistake than Wells, they made a hypothetical journey into the future to see what lies in store for us in 2050. They spend a day looking through the eyes of their imaginary 'heroine', Karen Rich. Textbox 13.1 is our shorter fantasy version of what they 'see'.

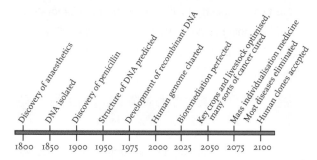

Figure 13.2. Timeline for health care and biotechnology, adapted from Shamel & Udis-Kessler (1999)

Francis Collins was also seduced into making predictions at the turn of the millennium (Potera, 2000). Using the unravelling of the human genome as his starting point, he even does it in quite a lot of detail for the first four decades of the 21st century.

Here are a few examples. He expected there to be a few successful cases of gene therapy in 2010. As we write, in mid-2010, his prediction has come true to a certain extent. In Chapter 11 we concluded that there had been some success with gene therapy, but that the awaited revolution still hadn't taken place. Collins also thought that serious discussions would take place on the broad application of diagnostics before fertilisation, in particular on the consequences thereof. This would be for the purposes of identifying genetic abnormalities that may result in disabled children. These predicted serious debates took place in the Netherlands in 2008, but at the time the subject was embryo selection. The result was preliminary, very limited legislation on embryo selection in the event of a genetic predisposition for breast cancer.

As for 2020, Collins predicted that various drugs will have been developed on the basis of genetic knowledge to treat, for example, diabetes and high blood pressure, and that sensitivity to drugs will be assessed before medicines are prescribed. In 2030 he expects that our "ageing" genes will have been mapped, and that clinical research on extending life will be taking place. By then the analysis of an individual genome (DNA passport) will cost no more than $1000 per person. This latter forecast now seems over-cautious, since we are virtually at that juncture already. For 2030 he also predicted the availability of computer models of the human cell for research, and the presence of groups opposing technology in general and modern biotechnology in particular that are avidly protesting against all these new developments.

Diary of Karen Rich; 1 April 2059.

7-8 a.m.:

Switched off the alarm clock and switched on the coffee machine with my voice, as usual.

Looked wistfully at my grandpa's giant mop of hair in the last photo I have of him and can just about remember the genomics breakthrough that made it possible to cure baldness.

Put on my 'NaturalRed', no-iron cotton clothes and drink a cup of breakfast that is scientifically tailored to my morning physiology.

Spray on a little perfume, also customised for my body - it smells of African orchids, and it's completely biosynthetic.

8 a.m. – 4 p.m.:

'Work' in Lifeco's park-style factory churning out biotech-related products made from renewable raw materials.

4 p.m. – 6 p.m.:

Visit my GP for an annual check-up.

Give him my MediChip, which he inserts into the fully automated, diagnostic and therapeutic CustomMed 321.

Don't think I have any health problems and the report concurs.

But just to be on the safe side, decide to wait a few minutes to let the 321 perform a full diagnosis on a fragment of tissue sample.

Have a DNA passport that links up to several of the 5000 known genetic abnormalities that increase the risk of a certain disease; these include intestinal cancer which my grandfather finally died of at a very late age.

Hardly feel the nano-needle go in as the 321 diagnosis begins.

In next to no time, I find out that there is a microscopic polyp in my intestines and also what the best prophylactic treatment is.

Decide to begin treatment right away.

Half an hour later and I'm on my way home after a microinjection from the 321 and with a medical cocktail in my pocket to drink before dinner.

Suddenly realise that 50 years ago there was a good chance that this polyp would have gone unnoticed until it was too late.

6 p.m. – 11 p.m.:

Arrive home and choose one of my 'personal diet TV meals' – carb-rich bread with a fat-free, protein-rich beef-flavoured tempeh cutlet, a crunchy, vitamin-E enriched salad, and for dessert a delicious chocolate mousse that lowers my cholesterol and regulates my insulin levels.

Sit back in my relax-and-massage chair to watch a recent holographic film and enjoy my meal.

11 p.m.:

While getting into bed, realise that I can't imagine how life must have been 50 years ago, in 2009 when people had so many health problems and other worries, like the credit crunch and Darwinists and Creationists hammering each other over the head.

Am grateful to the scientists who, with a little help from modern biotechnology, were able to make the world a healthier place.

Just before I nod off, I wonder what life will be like in another fifty years, in 2109.

Finally, Collins predicted that by 2040 health care will be much more extensive thanks to knowledge about our genome. Predisposition for diseases will then be established by looking at individual DNA passports and personal preventive health care will be available and effective. The testing of neonates for predisposition to diseases in later life will also be possible, although it will not yet be possible to take into account all the environmental factors. In addition, gene therapy and gene-based medicines will be available for most disorders and the average lifespan will be 90. International tension will increase due to socio-economic inequality in terms of access to medical treatments. There will also be debates on the classification of human traits and characteristics. This last reeks of eugenics, the extremely controversial theory that the human race can (or should) be improved by selecting individuals with 'desirable' characteristics (e.g. good health, beauty, intelligence, etc) before reproduction.

If we take all the predictions together, the general expectation seems to be that there will be a paradigm shift in health care in the course of the 21st century. At the moment we go to the doctor when we feel unwell. He makes a diagnosis and prescribes a treatment. The expectation now is that there will come a time in the 21st century when not only can an individually 'preventive service book' be drawn up on the basis of our personal DNA passport for serious and less serious 'events', but we will also be following a personalised 'preventive diet'. All with a view, of course, to a long and healthy life!

13.5. WILL THE NETHERLANDS CLIMB ON THE BANDWAGON?

The decoding of the human genome was, as mentioned above, a thrilling race between the entrepreneur Craig Venter on the one hand, and a consortium of mainly American and British scientists on the other. The rest of Europe, including the Netherlands, stood on the sidelines, but when Clinton and Blair presented the map of the human genome in a joint show with Collins and Venter, the Dutch government suddenly seemed to wake up. In mid-2000, at the government's request, industry and universities wrote the Strategic Action Plan for Genomics. This plan made recommendations for investment in DNA research. A year later a follow-up committee delivered concrete proposals and advised the cabinet to invest € 270 million in genomics over the next five years, on the basis that the investments should contribute to the improvement of the quality of life of the population. Investments should therefore be focused on the relationship between diet and health, methods for improving food safety, mechanisms of infectious diseases, the occurrence of disorders influenced by both genetic and environmental factors, and on the functioning of ecosystems and sustainability - focusing on environmentally friendly and healthy plant and animal products. Another interesting recommendation was to develop objective information so that individual members of the public could come to a more balanced view. Activities in those five years would prove whether enough had been done to catch up with the top groups in this promising area, where the requisite investment seems astronomical, but where the profit can be even greater if the above-mentioned expectations can in any way be realised. And after those five years it would also be demonstrated whether people had a more balanced view on these matters. One thing is sure, the Netherlands Genomics Initiative (NGI), which was set up in 2003, has put the Netherlands back on the genomics map. This initiative began a second five-year period in 2008 with the help of another € 280 million of government money[93]. Whether this has all helped to create a more balanced public opinion is less certain. We've certainly not been aware of it.

The Dutch contribution to the analysis of the human genome was therefore fairly minimal in the first draft versions. The person who was most involved was Gertjan van Ommen, head of the Department of Human Genetics at the Leiden University Medical Centre, and founder of the Leiden Genome Technology Center. He was chairman of the Human Genome Organisation, or HUGO[94], not to be confused with the Human Genome Project. HUGO had no direct role in the final phase of the Human Genome Project. Van Ommen: "As far as that was concerned, we at HUGO simply stood on the sidelines and applauded." HUGO is an ongoing concern with 1,200 members around the world and existed before the "sequencing" began. The organisation is primarily concerned with the dissemination of technologies, patents, ethical aspects, name allocation and gene mapping. During an interview in 2001, Van Ommen said the following

about the unexpectedly small number of human genes compared to the number of human proteins: "So the correlation between genes and proteins is really small, there's no debate about that now. We still know too little about the expression levels of DNA, RNA and protein to be able to say anything intelligent about the correlation between them. There are many preceding regulatory steps. With every publication I am again surprised about how complex the working of genes is. And this was not the only surprise that came with the unravelling of the genome."

13.6. THE SURPRISES OF THE GENOME

On the back of the above-mentioned publications in *Nature* and *Science*, the Dutch magazine *BIOnieuws* dedicated most of its edition of 3 March 2001 to the unravelling of the human genome. The "ten surprises of the genome" were also included in this edition:

1. *The genome is like a wasteland.* The genome is not a nicely arranged row of genes. Extensive "deserts" of millions of bases where nothing seems to happen, are alternated by densely populated "urban areas" of genes.
2. *We have a lot fewer genes than expected, approx. 31,000 (since recalculated at 23,000).* Earlier estimates started at 100,000 genes or more, but those figures were not based on the whole genome.
3. *The human genes can make three times more proteins per gene than the genes of a simple organism.* Human genes are often built up of little

[93] www.genomics.nl
[94] www.genenames.org

pieces of separate DNA, exons, which can encode different proteins in various combinations (Textbox 13.3).

4. *The architecture of many human proteins is more complex than that of a simple organism.* Many human proteins are multifunctional tools, with several active components - the *domains*.

5. *More than 200 human genes are the result of horizontal transfer of bacteria.* These homologues don't occur in fruit flies, roundworms or yeast. The genes appear to be good at self-preservation or even enter into a symbiotic relationship with the host.

6. *Self-replicating base sequences store about half of our DNA.* This DNA is a fossil archive that enables us to look back almost 800 million years in evolution.

7. *There is also 'junk' DNA which appears only to exist for its own sake.* For molecular biologists this is the irritating selfish DNA, such as the segment christened Alu. This piece of DNA, consisting of 200 to 300 bases, occurs a million times in the human genome.

8. *The mutation speed is two times faster in men than in women.* This means not only that the majority of mutations occur in men, but that the latter are also unconsciously responsible for evolutionary changes.

9. *There is no scientific basis for racial distinction.* At the DNA level all people are 99.9 % alike.

10. *The genome already offers benefits for health care, for medicine and for the development of medications.* Genes responsible for human diseases that had not yet been found, are now being discovered via known DNA data.

The most important outcome is a paradigm shift in our genome knowledge (Textbox 13.2). Knowing, for instance, how so few genes can encode such a complex organism; after all, humans have only approximately twice as many genes as, for example, a fruit fly, while worm, sand rocket and rat have almost the same number. Suddenly the concept of one gene, one protein becomes much too oversimplified, at least in the case of humans.

13.7. WHERE ARE WE NOW?

The ultimate goal of the Human Genome Project was to construct a map 'portraying' all human genes, i.e. mapping the entire sequence of all base pairs (the building blocks) of all the genetic material - the DNA - of humans. A large part of the sequencing was determined in 'draft' form for the publication in 2001. Careful verification was still necessary to rectify multiple mistakes and it was precisely these mistakes that rendered a correct sequencing analysis extremely complex. Even then there were still large gaps in the sequence that remained undetermined. Attention was primarily focused on gene-rich pieces of chromosomes,

TEXTBOX 13.2.

Paradigm shift: one gene = one protein → one gene = several proteins.

DNA contains information in the form of genes which tell a cell which proteins to make (Figure 13.3). Proteins regulate all the processes in our body. Proteins themselves are made up of smaller units, called amino acids (20 different kinds), which are present in the cell. In order to be able to make proteins, first of all a copy (transcription) is made of a small part of the DNA, a gene. This transcript is called messenger-RNA (mRNA) and is in fact an anti-copy, a sort of mirror image of the gene, that is to say mRNA is complementary to the gene. Both ends of the complementary DNA strings are different. One side ends with a phosphate group (see also Figure 1.1 in Textbox 1.1) and is called 5'-terminus. The other side that ends with the ribose pentagon is the 3'-terminus. Formation of mRNA can only start at the 3'-terminus, so the reading of the gene in the DNA string takes place in the direction from 3' to 5'. During the reading the DNA helix opens and closes as if it were a closed zipper with two fasteners that move at a close distance from each other in the same direction and open and close the zipper respectively. At the end of 2008 Nature published an article about a genetic analysis which appeared to show that not one but several proteins could be made from nearly all human genes. This occurs via an interim step, between gene and mRNA (Figure 13.4). First a temporary mRNA is made, then sections, introns,

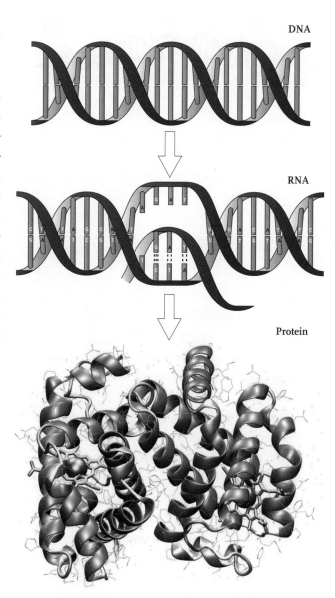

Figure 13.3. DNA → RNA → Protein.

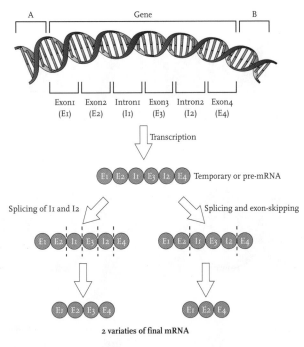

Figure 13.4. One gene is translated in several different ways into messenger-RNA, adapted from Van Santen (2008).

from that are cut out (spliced). An intron is a piece of the gene that does not contain code for the pertinent protein. The code for pieces (domains) of the protein can be found on the so-called exons. As the figure shows splicing can be done in various different ways. The mRNA can also be spliced at a different place (alternative polyadenylation). The net result is that sometimes a fragment of gene is passed over, while others have two different start or end sites, etc. This is how two or more active proteins can be created. A close-up of fragments of DNA around the genes (A and B) shows other codes which researchers suspect regulate the amounts of the various proteins. The proportions can vary dramatically between different tissues. This orchestration of these processes ensures that a heart becomes a heart, and a brain becomes a brain, etc.

the so-called euchromatic segments, while the gene-poor segments (heterochromatic) have received little attention so far. And yet these latter segments also contain important structural and regulating elements, for example the genes that play a key role in the earliest embryogenesis.

The first "draft versions" of our genome therefore still contained a great many gaps and uncertainties. Yet doctors, geneticists and pharmaceutical companies were able, for example, to start researching the genetic causes of some diseases or developing new therapies and medicines for them. When the Human Genome Project officially ended in 2004, our genome had still not been completely mapped. The burning question is whether a final version, in which 99.9% of the human genome has been determined with extreme accuracy, will ever come to pass. The relatively difficult and therefore expensive mapping of the last segments also provides little extra information, while the really challenging and essential work has really only just begun, i.e. the search for the significance and effect of all these pieces of DNA. How do you link a piece of DNA to a function in the body, to a disease, to an abnormality? And what happens if, because of a mutation, for example, there is an incorrect base pair in the sequence? Only when we have the answers to these sorts of questions can we make full use of the code of the human genome and can the above-mentioned predictions come true. The explanation and study of life processes based on the complete set of an organism's genetic information is called *genomics*. This is, therefore, the area where attention must be focused in the coming decades. Since the Human Genome Project ended, great progress has been made in answering some of these questions. Below are a few examples in brief.

During the race between the Human Genome Project and Celera Genomics, the researchers from both teams combined samples from various individuals, for reasons of time and money as well as privacy issues, to create a sort of "reference" encoding strand, so that they only needed to analyse half, i.e. 3 billion, of the bases. They did so on the assumption that little detail would be lost as a result, since the genetic variation between different genomes was estimated to be no more than 0.1 percent. However, an article published in 2007 suggested that this was incorrect. In that article the DNA sequence of both sets of chromosomes of one person was fully described. The genome was that of none other than the gene hunter Craig Venter himself. Together with colleagues from the J. Craig Venter Institute and three other universities they revealed in the October 2007 issue of *PLoS Biology* (Gross, 2007) the sequence of all 46 chromosomes, that is, the set from his father *and* his mother. A comparison of these two sets reveals that there are many major differences.

Other studies from that time also corroborate this. In addition, in May that same year, there was an announcement about the unravelling of the complete genome of another well-known person, i.e. James Watson - who together with Francis Crick won the Nobel Prize for Medicine for discovering the double DNA helix, which elicited Salvador Dali to say: 'The announcement of Watson and Crick about DNA ... is for me the real proof of the existence of God' (Stent, 1974). By the end of 2007 it had been established that one individual can easily differ from another individual, or from something like the above-mentioned average reference genome, by as many as 15 million of the 3 billion bases, i.e. not 0.1 but 0.5 % genetic variation. This discovery of the many major genetic differences between individual humans was hailed by *Science* as *the* scientific breakthrough of the year 2007.

As more genomes are unravelled, the question of protecting privacy will come further to the fore (Cohen, 2007). Watson, for example, didn't want the status of the key gene for Alzheimer disposition in him to be made known. Venter, in contrast, laid himself genetically bare. In the *PLoS Biology* article (Gross, 2007) there is even a list of more than 20 of his gene variants that are associated with an increased risk of alcoholism, anti-social behaviour, tobacco and other addictions, heart disease and Alzheimer's. Venter is very relaxed about making his genome public. He stresses that in the great majority of cases, genetic features and diseases are not determined by one single gene. The more people that have their complete genome, characteristics and health status revealed, the easier and more reliably scientists can interpret the still very enigmatic human genome (Textbox 13.3.). Venter: 'we have nothing to fear, and everything to gain!'

"A total gene passport is worth nothing", this is what Edwin Cuppen contended in October 2008 at his inaugural speech as a professor of Molecular Genetics in Utrecht, the Netherlands. He had had his own genes mapped for this event, and made them public during his oration, with the deliberate aim of provoking discussion. By his own admission he had to think twice before laying his whole "life" open. Especially since when you make your own genome public, you are also exposing half the genome of your children. Is that really what you want? And what if some unpleasant facts come to light in the process? Cuppen defended the proposition that, after weighing up all the pros and cons, a gene passport is a good development. Using apparatus in his own laboratory he had his own DNA, but not his whole genome, screened for a number of known genetic characteristics. In fact, this is comparable to what the fast-growing DTC (direct-to-consumer) market of genetic testing in America is offering: for a few hundred dollars you can be screened for a short or long series of genetic characteristics by companies with amusing names like 23andMe[96], deCODEme[97] or the less amusing, but cheap, Pathway Genomics[98]. Cuppen didn't need his whole genome mapped to make his point. What he aimed to do was to prevent the sort of atmosphere surrounding the introduction of GM crops from building up around DNA profiling. His proposition is that we should use these new techniques to our advantage. His response to the question of why we need such a gene passport was: "Doctors and insurers won't need it for many years to come. The new sequencing techniques are of principal importance in the clinical environment. You don't want a passport of a patient, but a passport of his

[95] www.personalgenomes.org

[96] https://www.23andme.com
[97] www.decodeme.com
[98] www.pathway.com

tumours is very useful. With this you can sometimes make a better estimate of whether a certain treatment will succeed or not. It is more efficient, may prevent a lot of discomfort, and is also cheaper in the end." We applaud such debates. They are vital if we are to have effective legislation, especially concerning privacy and insurability. In 2008 a law was introduced in the US prohibiting discrimination on the basis of genetic factors.

There is still no 1000-dollar genome as such, but it is on the way, requiring just a few finishing touches to existing methods. In 2007 it was Venter and Watson who had their entire genome unveiled (at a cost of 1 to 2 million dollars); in 2008 it was the turn of an unknown Chinese and an African. The analysis of their genomes, or at least 99% of the sequence and all the interesting bits of it, took two months and cost 250,000 dollars, but by the end of 2008 there was one company which claimed to be able to sequence a human genome for 5,000 dollars. The 1000-dollar genome will offer many possibilities for cancer treatment. Cancer is a real DNA disease. It is the changes in the DNA of body cells that convert a cell into a cancer cell, and it is these DNA changes that determine the behaviour of the tumour - how fast it grows, or spreads and what chemotherapy it is sensitive to. All this information is stored in the base sequence of the tumour cell DNA. Affordable and reliable access to this knowledge may therefore revolutionise the treatment of cancer. The 6 November 2008 issue of *Nature* contained the first ever description of the whole genome of a human cancer cell, as well as the above-mentioned genomes of the Chinese and

African. As the entire cancer genome was screened, mutations were found in unexpected places. According to Professor of Haematology Bob Löwenberg of the Erasmus Medical Centre in Rotterdam, these would have been missed if only the known gene regions had been examined.

If you stretch out all the DNA strands of the 23 pairs of chromosomes in a human cell and lay them end to end, they will measure two metres. How the cell manages to cram this into its minuscule nucleus is still an almost complete mystery. It's a bit like trying to fill a tennis ball with a twenty kilometre long rope. What we do know is that histones play a role here. Histones are proteins that enable the DNA strands to lie close to each other, without getting in a tangle and becoming completely inaccessible (Textbox 13.4). In 2008 another piece of this mystery was solved by researchers at the Dutch Cancer Institute and the Erasmus Medical Centre. Bas van Steensel and his colleagues wrote in *Nature* of 7 May 2008 that long pieces of human DNA strands are stuck to the inside of the cell nucleus wall (Figure 13.5).

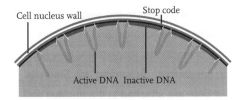

Figure 13.5. Cell nucleus: inactive pieces of DNA glued to the nucleus wall, adapted from Rouwé (2008).

The researchers found 1,300 marked pieces of DNA that stuck to the nucleus wall. These pieces were

Decade On, as the headline of the editorial in *The Lancet*, reviews the book *Drawing the Map of Life – Inside the Human Genome Project* by Victor K. McElheny[99]. It is written by Angela N.H. Creager in *Science* of 9 July 2010 and she concludes with: "The book's depiction of current trends in biomedicine, with the decline of 'gene-centered' accounts of traits and disease, seems less like a paradigm shift than a new frontier, once again driven by new technologies. The future trajectory, McElheny suggests, is promising though unpredictable. *Drawing the Map of Life* sketches out a more complete history of genomics than previously available, but clearly the story is not yet finished." This chapter is!

[99] shass.mit.edu/news/news-2010-mcelheny-drawing-map-life

13.9. SOURCES

Cohen, J. (2007). Genomics: Venter's genome sheds new light on human variation. *Science, 317*(5843), 1311-1311.

Collins, F. S. (2006). *The language of God: a scientist presents evidence for belief.* New York, Free Press.

Gross, L. (2007). A New Human Genome Sequence Paves the Way for Individualized Genomics. *Plos Biology, 5*(10), e266.

Hodges, C., Bintu, L., Lubkowska, L., Kashlev, M., & Bustamante, C. (2009). Nucleosomal Fluctuations Govern the Transcription Dynamics of RNA Polymerase II. *Science, 325*(5940), 626-628.

Noble, D. (2008). *The music of life: biology beyond genes.* Oxford University Press, USA.

Potera, C. (2000, August). Life after human genome map. *Genetic Engineering News.*

Rouwé, B. (2008, 10 May). DNA hangt met zijn inactieve delen aan de kernwand. *NRC.*

Shamel, R. E., & Udis-Kessler, A. (1999, December). Biotechnology in the 21st century. *Genetic Engineering News.*

Stent, G. (1974). Molecular biology and metaphysics. *Nature, 248*(5451), 779-781.

Sulston, J., & Ferry, G. (2002). *The common thread: A story of science, politics, ethics, and the human genome.* London, Joseph Henry Pr.

Van der Laan, S. (2000, 12 August). De geheimen van het genoom. *Chemisch2Weekblad,* p. 31.

Van Santen, H. (2008, 8 November). Veelzijdige boodschapper. *NRC.*

Venter, J. C. (2007). *A life decoded: my genome, my life.* Viking Press.

Zwart, H. (2008). Understanding the Human Genome Project: a biographical approach. *New Genetics and Society, 27*(4), 353-376.

STEM CELL THERAPY: PROMISING AND CONTROVERSIAL!

"Mankind has been forever in search of eternal youth. Where magicians and alchemists failed in their efforts, the biomedical scientist seems to offer the promise of eternal life with the discovery of the stem cell."

Hans Clevers and Ronald Plasterk

This is a quote from the foreword of a book called *Stamcellen* (Stem Cells), written by one of the world's leading stem cell researchers, Christine Mummery, and two of her colleagues (Mummery, Van de Stolpe, & Roelen, 2007). Stem cells are cells that, depending on the conditions, form specific cell types, tissues and organs. They don't yet have any specific or specialist function like normal (somatic) body cells, for example blood cells, skin cells and liver cells.

14.1. HUMAN EMBRYONIC STEM CELLS ARE 'HOT'

In the earliest stage of a human embryo, immediately after the first cleavages of the zygote (fertilised egg), all cells are still identical, still undifferentiated and can in principle still multiply endlessly and develop under the right conditions into any of the two hundred and ten differentiated, adult cell types that go to make up our body. This great capacity of the early embryonic stem cells to differentiate is called pluripotency. Pluripotency and the endless growth capacity are the characteristics that are useful in stem cell therapy: the targeted cultivation of a specific sort of tissue cell from embryonic stem cells in a laboratory, and the transplantation of these to a patient, for example heart cells to someone who has lost heart muscle tissue following a heart attack.

"Embryonic stem cells have achieved prominence in part because of the still unsubstantiated hopes that therapies that use them can ameliorate a variety of human ailments. They have attracted controversy mainly because the cells are obtained from human embryos, linking stem cell research to historical battles over abortion and over the legal and moral status of the human embryo and fetus." This abstract is from the chapter on stem cells in the book "The Art and Politics of Science" by Harold Varmus (2009). Varmus won the Nobel Prize for Medicine in 1989 and was director of the American National Institute of Health (NIH), which allocates billions of dollars every year to medical research in the US. His point of view, to which we subscribe, clearly represents the status of stem cell therapy in the first decade of the 21st century: promising yet controversial! Afraid of controversy, the Bush government largely restricted activity in this area, but fortunately, with the entry of Obama into the White House, clinical stem cell treatments look within reach again (see next section).

When the first edition of the book "Stamcellen" was published in 2006, Clevers and Plasterk were both directors of the Hubrecht Laboratory in Utrecht, the centre par excellence for research on developmental biology and stem cell therapy. It is an institute of the KNAW (Royal Netherlands Academy of Science), where Mummery worked for years. Since 1 April 2008, she has been a professor at LUMC (the Leiden University Medical Centre). Plasterk left science for politics in 2007, to become Minister of Education. Stem cell research obviously is a dynamic field, not only in the lab. Since the appearance of the stem cell book another spectacular and paradigm-shifting development has taken place, namely the manufacture of human pluripotent stem cells by the induced de-differentiation of specialised adult cells, or in other words the "reprogramming of differentiated normal cells back into pluripotent undifferentiated stem cells" (Section 14.7). There is a good chance that in the long run this will remove the need to make pluripotent stem cells from embryos, and thus avoid the accompanying controversy. Despite these rapid new developments, the book by Mummery et al. is still well worth reading, because it discusses in a clear way the underlying ideas and principles of stem cell therapy. The interested Dutch reader can use this as a basis for understanding and interpreting the new developments, discussions and publications on this subject in journals or newspapers, and thereby establish an informed opinion on this hot topic. An updated English version could in our opinion form a very useful contribution to educate the broader public globally. Anyway, we were very grateful for the use of the book *Stamcellen* to write this chapter.

14.2. FROM BUSH TO OBAMA

Shortly after President Obama was inaugurated in January 2009, the Californian company Geron was given the go-ahead by the FDA to start the first clinical trials with human embryonic stem cells. Although the FDA vigorously denied that this had anything to do with the arrival of Obama in the White House, nonetheless it did mark the beginning of a new era. Obama's predecessor, President George Bush jr, had in 2001 restricted all federal support for research with human embryonic stem cells to 15 then-existing, NIH-registered, stem-cell lines, the "Presidential list". More indications of a new stem cell era quickly followed. In March 2009 Obama declared that he wanted to lift the restrictions that Bush had imposed. A month later, the Obama government laid down guidelines to regulate this research. In mid-2009 a White House spokesman announced that the administration was busy processing approximately 50,000 reactions from the public on these draft guidelines and that the final guidelines would be ready by early July 2009. These

would, according to him, form the framework within which research on embryonic stem cells could evolve in a scientifically valuable and responsible way. However, Tom Okarma, head of Geron, said that the draft guidelines showed that government officials have still not fully understood the potential of human embryonic stem cells. Given the potential of stem cell therapy, a much broader framework and much stronger incentives are required. In his view, stem cell treatments will be able to cure previously untreatable diseases, thereby saving lives as well as money that is currently spent on ineffective medicines. Treatments with embryonic stem cells are not 100,000 dollar therapies that extend lives by three weeks, says Okarma. With a simple intervention you can permanently repair a defective

FROM BUSH TO OBAMA ...

I - 0 FOR ME, MISTER BUSH!

function of an organ or tissue that has been damaged by an injury or disease. According to Okarma what is needed is a presidential committee of experts to advise the administration on government policy, allocate research grants, and promote collaboration between researchers in industry and academia.

The final guidelines for allocating federal money to stimulate stem cell research, more than ten billion dollars, were made public on 6 July 2009. Only research on stem cells from embryos left over after *in-vitro* fertilisation (IVF) is eligible for federal funding, provided there is written consent from the donors. How all this will evolve is not yet clear, but Geron began the first clinical trials in early 2009. The trials involve experimental treatment with embryonic stem cells for patients paralysed by transverse myelitis (spiral-cord injuries). The study was however halted after seven months because safety concerns surfaced in an animal study, which showed an increased frequency of small cysts within the injury site in the spinal cord. In response, Geron developed new testing methods that essentially ensure the purity of the drug, which is actually a mix of different cells. On 31 July 2010 Thomas Gryta writes in the online *Wall Street Journal* that the FDA has cleared Geron to resume their stem cell study. The initial testing in humans will focus on the safety of the drug, and its effectiveness must still be proved. The study will evaluate safety in eight to ten patients with recent severe spinal-cord injuries. The company agreed with the FDA to leave 30-day intervals between the first patients, for safety reasons.

Researchers will monitor the patients for over a year to find out whether the treatment is safe and whether defective functions and movement possibilities have been repaired.

If this clinical trial is successful, much of the resistance to applications with embryonic stem cells will probably fall by the wayside. In any case, the new era has started with an explosion of spectacular novelties. There isn't a day goes by without one or two appearing on the Internet. On 27 July 2009, for example, there was a report on the identification of the most suitable stem cells for cultivating bone implants and a report on surgical thread with embedded own stem cells to stimulate the healing process of sutured wounds. A year later, as another example, in the July-August 2010 issue of *Euro/Biotech/News* it is reported that Italian researchers have restored sight to blind patients using stem cells from the patients' own bodies; it concerned 106 patients whose eyes had been severely damaged by chemical burns. The preceding May-June issue of this journal brought somewhat unexpected, but pleasantly surprising news: "The Vatican has taken a bold step into unchartered territory with its decision to finance new research into the potential use of adult stem cells for the treatment of intestinal disease and possibly other conditions." Although the Vatican has a positive stance with respect to transgenic crops (see Textbox 3.1 in Chapter 3) and to biotechnology in general, it is on the understanding that there should continue to be a ban on cloning humans and tinkering with human DNA (Chapter 1). Nevertheless this news seems to be a first step in a slight loosening of the ban.

The current debates on stem cells and the political policy that regulates their use, have been influenced by several crucial events, whereby the position of successive American governments has had a major impact on what happened in the rest of the world. Each of these events was of critical significance for the course of stem cell research. It began in 1978 with the birth of Louise Brown, the first IVF baby. *In-vitro* fertilisation is the fertilisation of human eggs in the laboratory and, after a few cell divisions, the implantation of the formed embryo into the uterus, giving childless couples a greater chance to have a child. IVF took off after the birth of Louise Brown. Since more eggs were fertilised than embryos implanted, and there was no possibility of preserving the excess IVF embryos for later IVF use, there was little discussion about their alternative use. As a consequence and with the minimum of discussion, in Oxford (England) amongst other places, the remaining embryos were used in the first, albeit failed, attempts to obtain an embryonic stem cell line. A cell line consists of the same type of cells which can be further cultivated under suitable conditions in the laboratory. Only when, a short while later, a freezing procedure for embryos was developed, did the ethical discussions on the use of remaining IVF embryos begin, particularly because the stock of frozen IVF embryos grew exponentially. The often heated discussions resulted in the blocking of research on human embryonic stem cells, and research shifted again to the stem cells of mice.

It wasn't until 1994 that interest in human stem cells was renewed. In that year an NIH (American National Institutes of Health) panel issued a report on possible prospects for research into early embryogenesis. It was written with the prospect of there being a new, and in all expectation, permissive political policy. The recommendations for research on human embryos in this 1994 report were almost identical to recent and promising work at that time on mouse embryos and stem cells. The report also expressed the expectation that major progress in mammal biology would result, which would greatly facilitate the successful application of human embryonic stem cells in clinical research. Changes in the political climate, however, led subsequently and all too quickly to a ban which suspended much of the research recommended in the report. In 1996 US congress banned the use of federal funds to create or destroy human embryos solely for research purposes.

It was Ariff Bongso, a pioneer in the field of stem cell research, who was the first to recommence work on human stem cells in Singapore. In 1994 he described a procedure for removing cells from a blastocyst and cultivating them in a petri dish in the laboratory (Section 14.4). A blastocyst (Figure 14.1) is a small round structure filled with fluid and cells, which is formed after several divisions of the fertilised egg. In mammals, where implantation of the blastocyst takes place in the uterus, the cells which form the actual embryo are located as a cluster of cells, the *inner cell mass* (or embryoblast), eccentrically in the blastocyst.

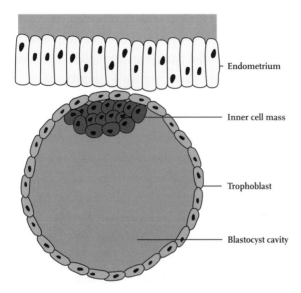

Endometrium

Inner cell mass

Trophoblast

Blastocyst cavity

Figure 14.1. A blastocyst at the endometrium, the mucosa that lines the inner wall of the uterus in which the blastocyst becomes implanted.

The American James Thomson used this method in 1998 to make the first human embryonic stem cell line. This scientific breakthrough breathed new life into stem cell research and in no time various laboratories were isolating new human stem cell lines. This research has been politically and ethically charged from the beginning, and boycotted to a large extent by the Bush government. As previously mentioned, in 2001 President Bush banned government funded research with human embryonic stem cell lines which were made after 9 August 2001. This effectively restricted stem cell research in the US thereafter to the 'presidential list', the fifteen stem cell lines which were officially registered at the NIH; these cell lines were all difficult to cultivate and therefore not really suitable for research or use in applications.

For a long time the use of human embryos for the creation of stem cell lines was only possible in a very limited sense in many European countries too. Yet, a number of research groups inside and outside the United States, and especially in Asia, stubbornly persisted with the development of new human embryonic stem cell lines and made significant advances. The International Stem Cell Initiative was set up in January 2003, and decided in 2005 to compare all registered human embryonic stem cell lines with each other to establish similarities and differences and to stimulate further research; 75 cell lines from 14 countries around the world were involved in this research. Since 2002 the use of human embryos for making stem cell lines has been permitted in the Netherlands under certain conditions (with the consent of the CCMO (Central Committee on Research Involving Human Subjects)). A declaration of consent from the donors of the embryos (both the man and the woman) is a prerequisite. Since July 2009 that has also become a crucial precondition in the US. However, because of the restrictions many research groups have shifted their attention to less emotionally charged research on adult stem cells. Stem cells occur not only in embryonic tissue, but in virtually all our tissues. Admittedly, these adult stem cells have more limited differentiation possibilities, but this field has nevertheless advanced in leaps and bounds. After years of fundamental research, these adult stem cells are now beginning to bear fruit in medical applications[100]. In December 2008 the Translational Adult Stem Cell research programme began in the Netherlands. It had a budget of more than 22 million euros for research into making stem cell therapy a reality for patient care.

The final crucial event was the birth of Dolly the Sheep in 1997 - a remarkable scientific achievement. The way in which biologists looked at the arrangement of genetic information changed fundamentally as a result. Up till then it was thought that the transition from undifferentiated stem cells to fully differentiated, specialised tissue cells was effectively irreversible. All body cells do have precisely the same genome, but in stem cells there is a completely different package of active genes from those in specialised cells. After many futile attempts, the introduction of the genetic material from an udder cell from the "mother sheep" into an oocyte of another ewe and the implantation of the formed "embryo" in the uterus of a third, resulted in the clone Dolly. It showed that, in contrast to expectations, genetic reprogramming of adult specialised cells to much earlier stages in their development is a very real possibility. It also suddenly opened the way to 'patient-specific' stem cells for personal therapy. But Dolly's birth also raised fears of human reproductive cloning, which seriously limited any desire to design a promising method for reprogramming cells for therapeutic purposes. So it was more than ten years before the publication of the first examples of genetic reprogramming of specialised human cells back to the stem cell stage. By the end of 2007, it had happened: the era of the "formation of induced human embryonic pluripotent stem cells by dedifferentiation of specialised cells" had arrived, and led to feverish new developments without the ethically-charged label carried by stem cell lines isolated from human embryos (Section 14.6). It went hand in hand, in the US too, with a relaxing of the restrictions under which human embryonic stem cell research could be carried out. For example, at the end of 2004 a referendum in California resulted in the release of three billion dollars for embryonic stem cell research. The Californian Institute of Regenerative Medicine was set up, but the money only became available in mid-2009 after delays caused by lengthy legal procedures. Within four years, from the beginning of December 2009, the institute hopes to have ten to twelve new stem cell therapies in the clinical trial phase with humans. The aim of the institute is to promote the transition from tests performed on animals in the laboratory to tests in humans in the clinic, very like the Dutch Translational Adult Stem Cell Research programme. It looks therefore as if, in the early days of the Obama era, the lines have been redrawn. Time will provide the answers to the questions whether stem cells are ultimately suitable for therapeutic applications and for which disorders, but primarily with which stem cells: embryonic, adult, or induced pluripotent stem cells, or perhaps all three?

[100] www.xcell-center.com

14.4. WHAT IS A STEM CELL (THERAPY)?

The term "stem cells" has a well-defined meaning for biologists, and implies more than just the controversial, politically-charged types originating from human embryos. All specialised cells such as skin cells, liver cells and brain cells are formed in humans and animals by means of an orderly process, in which undifferentiated cells divide and differentiate into these specialised cells. Under certain conditions, the undifferentiated cells can form two types of daughter cells when they divide: a daughter cell that cannot be differentiated from the mother cell and does the same thing, and another daughter cell that moves towards becoming the specialised cell. These undifferentiated cells, which not only endlessly replicate but can also produce differentiated offspring, are called stem cells. Many stem cells also occur in adult tissue and have a limited ability to differentiate, for example, blood stem cells differentiating into all sorts of blood cells such as red and white blood cells, but not into insulin-producing cells. Such "multipotent" adult stem cells can divide in two ways (Figure 14.2).

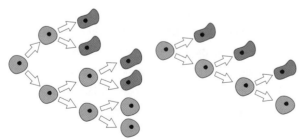

Figure 14.2. The two ways in which adult stem cells can divide, adapted from Mummery et al.(2007).

As we have seen, stem cells in a much earlier stage of human and animal development, i.e. in the early embryo, have much greater potential. These early embryonic stem cells are the progenitors of all 210 (differentiated) cell types from which tissues and organs of the adult human are made. It is because of this "plural" potential that they are called pluripotent. In principle, pluripotent stem cells offer the most possibilities for stem cell therapy.

Stem cell research is to the first decade of the 21st century what recombinant DNA research was to the 1970s and 1980s, and the Human Genome Project was to the 1990s - the most visible and most striking manifestation of the promising and spectacular developments in the biological sciences. Most stem cell biologists agree that human embryonic stem cells have the greatest potential in principle to treat human diseases and wounds (Gruen & Grabel, 2006). This expectation is based on the observation that these stem cells can differentiate themselves into most, but not all, cell types from which the adult human body is composed, not only in the body (*in vivo*), but also under suitable conditions *in vitro* (test-tube, Petri dish, etc.). The road to successful stem cell therapies therefore seems to be a straightforward one: make the desired specialised cell type from human embryonic stem cells, for example, pancreas β-cells for the treatment of diabetes type 1 (an autoimmune disease whereby the patient cannot make insulin in the body), and transplant these cells to the desired location in the patient. This is easier said than done, though, because the knowledge and technology required for such a process has not yet been sufficiently developed. There is an abundance of scientific literature on differentiation and transplantation of mouse stem cells, but the literature on differentiation and transplantation of human embryonic stem cells is lagging far behind. However, the distance between the two is diminishing all the time. Nonetheless, a great many hurdles, both ethical and scientific, still need to be overcome before we see the routine application of stem cell therapy in the clinic.

One hurdle that is not often explicitly mentioned is the resistance, not to say aversion, which a great many researchers have to investing time in acquiring a better understanding of the nature of possible ethical resistance to their work. It is usually these researchers that are best suited to feeding the ethical and political debates with relevant and objective information, providing that they have the right skills to (dare to) discuss and defend their work on ethical grounds. In our view, universities could play an important role here by giving the subject more attention and form.

An important scientific obstacle yet to be tackled is the risk of tumour formation, which accompanies every transplantation with undifferentiated cells. Many methods are being researched to eliminate the tumour-forming cells, as far as possible, before and after transplantation. As with any experimental therapy, an acceptable level of risk must be carefully defined. What is essential here is that the patients eligible for treatment are given as much information as possible on all aspects of their treatment. Until recently, nearly all human stem cell lines were isolated and cultivated in the presence of animal components, often in the form of serum, and/or additives and growth factors. Cells to be transplanted must be guaranteed to be free of animal substances which contain pathogens or cause immune reactions. For that reason, researchers are eagerly looking for effective purification methods for the existing cell lines and investigating isolation techniques and cultivating methods that don't use animal components. The risk of transplanted cells themselves being rejected is also very real. Since cells that are genetically identical to those of the patient are the most promising approach to this problem, intensive research is under way to find methods for making these "patient-specific" cells. Finally, there is an obstacle that has become much greater in recent times, namely the funding of stem cell research. Barack Obama has opened the doors for this research again and hopefully where he leads others will follow. It will certainly help scientists to overcome the other obstacles.

14.5. TYPES OF STEM CELLS

An investigation of the development of a fertilised egg into an adult human presents a good picture of the various different stem cells that exist (Figure 14.3). The merging of a sperm and oocyte creates a zygote, whereby the sperm cell is "swallowed up" by the much bigger ovum. A male and female set of chromosomes come together and fuse into one cell nucleus which contains the whole genome, i.e. in the case of a human cell nucleus 23 pairs of chromosomes. The cytoplasm contains the necessary components for this fusion, but also for the first cell divisions. Approximately one day after fertilisation the human zygote cleaves itself into a two-cell embryo, then into a four- and eight-cell embryo (cleaving divisions), whereby the original cytoplasm is distributed over all the cells, so that there is no change in the total volume. The cells of these early embryos are called blastomeres and in theory each is able, individually, to form a complete embryo that can nestle into the lining of the uterus and develop into a complete individual. This property is called totipotency.

After about three days and three cell divisions a solid cluster of eight blastomeres is formed, called morula (mulberry). From this stage onwards, the first morphological (shape and composition) differences between the cells can be observed and the totipotency rapidly decreases. After four days the blastocyst stage is reached, and more than a day later the blastocyst (Figure 14.1) implants itself into the uterus. There are only two cell types in the early blastocyst, namely the

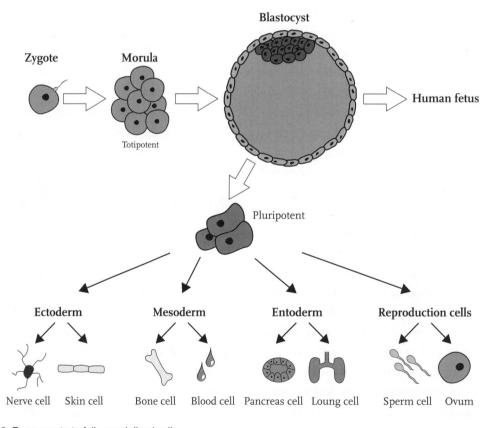

Figure 14.3. From zygote to fully specialised cell.

trophoblasts and those in the eccentric cluster of cells, the inner cell mass. The latter is the actual embryonic part which will later form the fetus and the new individual. The so-called extra-embryonic structures, such as the placenta and the umbilical cord, are not formed from the cells of the inner cell mass, but come partly from the trophoblast cells which surround the fluid-filled cavity and partly from the eccentric cluster of cells of the blastocyst. Since all cells of an adult human, except those of the extra-embryonic structures, can be created from the inner cell mass, they are called pluripotent.

The pluripotent stage doesn't last long, because the cells rapidly differentiate during the normal embryonic development into more specialised cells. In addition to the progenitors of the reproductive cells, three new cell types emerge from the inner cell mass. These are called cotyledons: the outer layer or ectoderm, the middle layer or mesoderm and the inner layer or endoderm. All human organs and tissues stem from these. A few examples of specialised cells that are formed from the various cotyledons are given in Figure 14.3. Stem cells are still present after the embryonic stage, but they have a more limited potential. In most, and perhaps even in all, organs and tissues of an adult individual there is a small stock of adult stem cells. Such organ or tissue stem cells can still divide and generally develop into a limited number of cell types of which the tissue or organ in question consists: they are multipotent.

Differentiating multipotent stem cells give rise to progenitor cells which can still divide, but don't yet have all the properties of the final fully matured cell. These progenitor cells are called unipotent, because they can only differentiate into one cell type. Yet they might still be interesting for cell transplantation, because in principle they can still multiply outside the body, albeit to a limited extent. The final completely differentiated cell has to exercise its role within the organ and can generally not divide or only to a limited extent. However perfectly it functions, such a cell is no longer suitable for cell transplantation purposes.

14.6. THE MAKING OF HUMAN (EMBRYONIC) STEM CELL LINES

The making of a cell line involves the isolation of a certain type of cell from a tissue or organ and the cultivation thereof, so that only this type of cell appears in the culture. A considerable number of cells can be cultivated from this cell line and frozen in small portions at very low temperatures. These frozen cells constitute the cell bank with which further work can be done over a long period of time. The standard procedure for isolating stem cells uses embryos in the blastocyst stage. The cells of the inner cell mass are isolated from the blastocyst and spread out on a special medium in Petri dishes. The development of a suitable medium on which the stem cells can multiply without differentiating, requires specialist knowledge, expertise and research. It is this approach (Figure 14.4A) that raises the ethical objection, i.e. that embryos are lost in the process.

A second method (Figure 14.4B), to which there is no objection in principle, involves isolating a single cell, a

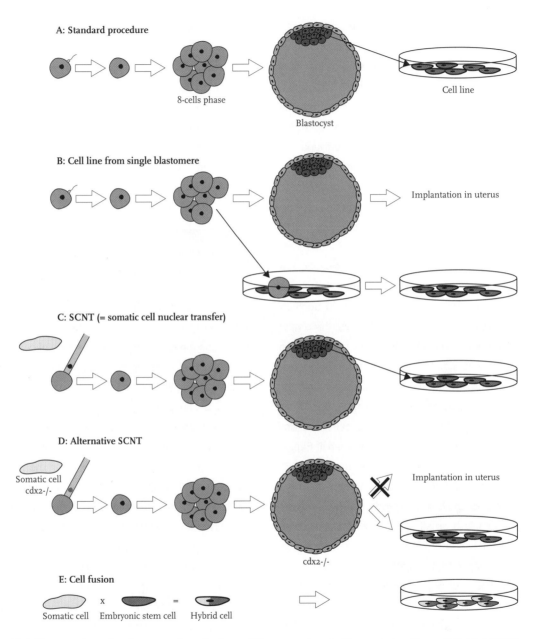

Figure 14.4. Five ways to make a human embryonic stem-cell line, reproduced with permission (Gruen & Grabel, 2006).

blastomere, from an even earlier stage of the embryo. The cell in question is one of the eight fourth-generation cells ($1 \rightarrow 2 \rightarrow 4 \rightarrow 8$) which come into being as a result of 3 cleaving divisions. The isolated blastomere is spread (plated out) on a medium which already contains a cell layer. Normally speaking many cells are needed to start the process of developing an embryonic stem cell line. By plating it out on an existing cell layer, it is possible to start from one cell. It is a well-known fact in animal cell culture that a minimum number of cells are necessary for growth. This cell layer, the so-called *feeder*, gives off a factor which prevents the cells from differentiating and losing their pluripotency; in 1988 this factor was identified as *leukaemia inhibitory factor* (LIF). When a blastomere multiplies on such a medium, the "adapted" daughter cells are isolated and further

---

TEXTBOX 14.1.
Pre-implantation genetic diagnostics.

Dutch embryo legislation requires that embryos left over from IVF treatment can be used under certain conditions for medical scientific research. However, they must not be specially created for this purpose. Through research and debate the Rathenau Institute[101] stimulates the public to make a judgement about scientific and technological developments. In 2008 this institute investigated the views of the public on the medical scientific use of both left over embryos and specially created embryos. This matter of embryo use became a sensitive subject later that year. In mid-2008, after the Secretary of State for Public Health had given the Maastricht University Medical Centre (MUMC+) permission to carry out embryo selection among carriers of a congenital breast cancer gene, there emerged a serious conflict in the cabinet. In the MUMC+ research included looking for serious genetic abnormalities in embryos before they were implanted in an IVF treatment.

The patients involved were not infertile, but had a life-threatening disorder in the family. Embryos with such a congenital anomaly are not implanted in the uterus. The intense political debate that resulted from this authorisation by the Secretary of State, surprised the researchers. They thought that embryo selection for congenital abnormalities – in this case for the breast cancer gene – would be a good alternative to terminating the pregnancy after an amniocentesis with an unfavourable outcome for the parents. For the researchers, abortion seemed morally more difficult to justify than embryo selection, for which an embryo had to be created. In the end, the cabinet decided that the embryo selection could proceed if the existing multidisciplinary committee in the MUMC+ continued to carefully examine each individual case. The MUMC+ also examines the seriousness and nature of the disease and the treatment possibilities, and has to submit new diseases that they want to present for pre-implantation genetic diagnostics (PGD) to a national guidelines committee on PGD (Rathenau Institute Annual Report 2008).

[101] www.rathenau.nl/en.html

cultured as a separate cell line. Human embryonic stem cell lines obtained in this manner have the necessary pluripotent characteristics, both *in vitro* and *in vivo*.

This second approach uses the remarkable regulatory capacity of the early mammalian embryo: if one or two cells are missing, it can regenerate the missing cells and form a complete embryo again. The goal is to isolate just one cell of the eight and make a cell line of it. The rest of the embryo can then be transplanted directly into the uterus, since it is still able to implant itself there and generate a complete embryo, fetus and finally a neonate. It can also be frozen for later transplantation. In fact, this method is no different from the one in which a blastomere is isolated for genetic pre-implantation diagnostics (Textbox 14.1). This approach also evokes the tantalising futuristic scenario in which IVF babies have a genetically compatible embryonic stem cell line in the freezer, which can be used later if necessary for all kinds of stem cell therapies without risk of rejection reactions. We certainly haven't heard the last word on this.

The third approach, which until 2006 attracted the most attention in terms of preventing immune reactions, involves transferring the nucleus of an adult somatic cell, for example a skin cell of the patient to be treated, to an oocyte from which the genetic material has been extracted with a micropipette (Figure14.4C); the extracted genetic material concerns one set of chromosomes, because the unfertilised oocyte is haploid (at this stage there is also no question of a real cell nucleus with a nuclear membrane). This approach is called SCNT or Somatic Cell Nuclear Transfer, or even therapeutic cloning. The genetic material to be transferred is usually obtained by performing a skin biopsy on the patient. The many cells contained in this piece of skin are allowed to multiply in a culture medium and are later detached from one another and from the culture medium using the enzyme trypsin. One of the cells is sucked up in a micropipette and injected between the cytoplasm and the surrounding *zona pellucida* (protective glycoprotein membrane surrounding the oocyte and the early embryo). The cells are fused using an electrical pulse and the nucleus containing DNA enters the cytoplasm of the oocyte. After a few more procedures, the resulting zygote is grown *in vitro* in a blastocyst and a human stem cell line is developed from this as shown in Figure 14.4A. The cells in this line are theoretically genetically identical to the cell from the nucleus and thus identical to the cells from the patient, if the donor cell came from him or her.

In 2004 and 2005 the first articles claiming success with SCNT were published in leading journals by the South Korean research group led by Hwang. However, in 2006, these claims appeared to be fraudulent. This was a major setback for human SCNT and tempered the optimism that had built up around it. In mid-2009, SCNT had still not resulted in human embryonic stem cell lines. Scientists continue, untiringly, with their attempts and there have been a few minor successes (Textbox 14.2). The ethical sticking point in this method is that blastocysts made in this way can be used not only for therapeutic cloning, but also for reproductive cloning. If such a

In the Netherlands and a great many other countries, cloning of people is forbidden. In Belgium, China, Spain and the UK, however, there are some institutes working in this area. In the US, too, cloning is authorised, but receives no federal funding, and it is therefore mainly industry that is carrying out research in this particular domain. Two competing American biotechnology companies have cloned human embryos with the aim of developing embryonic stem cell lines. Stemagen, in California, was first in early 2008. In the journal "Stem Cells" they wrote that they had made one cloned human embryo with donor DNA from an adult. With extensive genetic controls they also demonstrated that the clone really was a clone. After the Hwang disaster that was an absolute imperative. Despite the fact that it is impressive work, it appeared in a low-key journal and barely caught the world's attention. A year later the competitor Advanced Cell Technology (ACT) from Massachusetts published a similar study in "Cloning and Stem Cells". They had made 19 embryos that survived until they had cleaved three or four times, i.e. they consisted of 8 or 16 cells. ACT also properly verified that the clones were from an adult human.

During the transition from an undifferentiated embryonic stem cell to a final differentiated adult cell, all kinds of genes are deactivated and others just activated. For a successful cloning it must also be possible to reverse this process. ACT found proof that the gene activity of the cloned embryos does indeed resemble that of a normal human embryo and then we're talking about more than 5000 genes. However, no stem cell lines were manufactured from them. ACT did try, but found that with further cultivating more abnormalities appeared. The blastocyst stage was not reached and the standard procedure for generating cell lines (Figure 14.4A) could therefore not be used.

FURTHER CULTIVATION OF CLONED
EMBRYOS REVEALED ABNORMALITIES

MMM... THEIR SHAPE
AND COLOR ARE CHANGED!

The procedure used to make these human clones was the same one used to make Dolly. There were more than 200 failed attempts to get to Dolly; at the cost of many ova. The chance of success was, and still is, very small. This certainly applies to the cloning of donor DNA from adults. Human ova are difficult to acquire. In recent years there have been many heated discussions about an alternative way to clone human embryos, for example the use of animal rather than human ova. In mid-2008 the British Houses of Parliament voted to legalise these hybrid embryos and as such led the way in this area. However, ACT also blastocyst were implanted in a uterus, in theory it could develop into a full-blown individual, and, what's more, a 100 percent genetic clone of the DNA donor. The research is still at the theoretical stage for humans, but in mice it was already a reality in 2009 (Cyranoski, 2009). Tiny, the first cloned mouse, came about as a result of the reprogramming of a connective tissue cell of her clone parent. And Tiny is no longer the only one. Since Tiny, 27 other cloned mice have been "born" in this way. One of the males has since created healthy offspring, born after a normal copulation. The "reprogrammed clone premiers" live in Beijing and were created by researchers at the zoological institute of the Chinese Academy of Science. The research report of the Chinese scientists was published in *Nature* (Zhao *et al.*, 2009).

The fourth approach, the idea for which was launched by Hurlbut (Hurlbut, 2005), is in our view based on a dubious premise. The idea starts with a nucleus containing a gene with a mutation brought about by genetic modification; the pertinent gene is essential for the embryo to develop in the uterus. The sacrifice of such 'defective' embryos for the development of cell lines should be easier for some opponents to accept, because the embryo is not viable as a human being, and thus cannot be deprived of an existence by the creation of cell lines. The zygote made by nuclear transfer can cleave *in vitro* and produce cell masses for the blastocyst, and therefore cell lines, but the (induced) genetic defect prevents development in the uterus. The example given in Figure 14.4D concerns a gene that is essential for the implanting of the embryo in the uterus. Hurlbut is a professor at Stanford University and has a medical, ethical-medical and theological background. He served for eight years on the President's Council on Bioethics. So he's not a newcomer suggesting this dubious alternative[102]. He is at least thinking professionally about these difficult and ethically controversial matters (Glaser & Hurlbut, 2005). Induced pluripotent stem cells (next section) fortunately may make such an alternative unnecessary. The fifth protocol named in Gruen & Grabel (Gruen & Grabel, 2006) avoids the use of human oocytes and embryos to develop genetically compatible human

[102] med.stanford.edu/profiles/frdActionServlet?choiceId=showPublication&pubid=234636&fid=7484&

embryonic stem cells by fusing a cell from an existing human embryonic stem cell line with an adult somatic cell (Figure 14.4E). The chromosomes of the original embryonic stem cells must then be removed, so that the cells only have the chromosomes of the somatic cell, and thus of the patient. This is necessary for two reasons. Firstly, the chromosome complement of these hybrids is not stable in time. Secondly, if these cells preserve the DNA of the stem cell line, they are not genetically compatible with the patient. The technology to remove all the embryonic stem cell chromosomes is not yet available, will probably be very difficult to produce and is very labour-intensive. In addition, the removal has to occur after the reprogramming of the donor cell DNA, so that the hybrid cell has accepted the characteristics of a stem cell. A method for doing that has, however, not yet been established. Development and testing of this technology will no doubt take years and may well be as difficult and expensive as SCNT.

The technology published in 2007 regarding the induced reprogramming of adult human cells into pluripotent cells, places all these approaches in a very different light and probably makes them largely irrelevant (Baker, 2009).

14.7. FORMATION OF INDUCED HUMAN EMBRYONIC PLURIPOTENT STEM CELLS BY DEDIFFERENTIATION

In August 2006 two Japanese researchers from Kyoto University (Takahashi & Yamanaka, 2006) published a sensational article regarded by many fellow experts as implausible. By inserting four specific embryonic genes into the genome of specialised mouse cells, these were reprogrammed into cells that could effectively differentiate into any other body cell. In other words, specialised cells apparently possess enough plasticity and can be returned with relatively straightforward procedures to the pluripotent stage. The disbelief with which the results were received only spurred on Yamanaka's research group to generate more convincing proof. This was delivered ten months later (Okita, Ichisaka, & Yamanaka, 2007) in an article demonstrating that specialised cells could be reprogrammed into pluripotent stem cells, which could then differentiate into any specialised body cell. To their slightly unpleasant surprise, Yamanaka's research group had to admit that they were no longer the first with this proof, since two other laboratories published an article at virtually the same time claiming that they too had managed this (Maherali *et al.*, 2007; Wernig *et al.*, 2007). So began a heated race that was still in full swing during the writing of this chapter (mid-2010). In March 2009 alone, four articles were published describing a refinement of the technique. This fierce competitiveness is understandable: induced pluripotent stem cells are almost as promising as human embryonic stem cells, but without the ethical objections.

The first embryonic stem cells from mice were isolated in 1981. It wasn't until 1998, however, that the same was achieved with human embryonic stem cells. The time between the first induced pluripotent mice stem cells and those of humans is substantially shorter. By the end of 2006 pluripotent stem cells (iPS cells) had been induced from mice. By the end of 2007, and early 2008 there were human iPS cells (Park, Zhao *et al.*, 2008; Takahashi *et al.*, 2007; Yu *et al.*, 2007). Induced patient-specific pluripotent stem cells from patients with diabetes, Huntington's disease and muscular dystrophy are described in two articles, which were published in August 2008 (Dimos *et al.*, 2008; Park, Arora *et al.*, 2008); there are still none from embryonic stem cells for people with these diseases. According to Jeanne Loring, director of the Center for Regenerative Medicine at the Scripps Research Institute in La Jolla, California, the field of stem cell research is in danger of losing sight of the big questions because of the competitiveness. Questions such as: what are the mechanisms of reprogramming and what exactly will reprogrammed cells be able to do on a therapeutic level? She concludes the following: "Making cells is not the end point!" On the contrary, we the authors believe it is only the beginning of the biological challenges - the real therapeutic stem cell work.

On 19 July 2010 Carolyn Y. Johnson of the Globe Staff writes in The Boston Globe that the breakthrough discovery that scientists could transform adult cells into stem cells has sparked research in labs across the world, spawned start-up companies, and bolstered the long-term dream that a patient's own cells could be used to regenerate damaged tissue. Meanwhile, scientists have found that these cells, while similar in many ways to embryonic stem cells, contain subtle differences that affect their biology and therapeutic potential. Now, researchers all over the world are racing to understand the true nature and utility of the induced pluripotent stem cells (iPS cells).

Konrad Hochedlinger, a principal faculty member of the Harvard Stem Cell Institute, published with colleagues in the April 2010 issue of *Nature* (Stadtfeld *et al.*, 2010) that in most mouse iPS cells a cluster of genes, known to be important in development, was not activated. He found a small portion of iPS cells in which those genes were active, and the cells had the full development potential of embryonic stem cells. Konrad is now repeating the experiment with human cells, and says his work suggests that it may be possible to optimise the reprogramming process or to use the genetic differences to sort good iPS cells from bad.

In February 2010, researchers from Stanford University School of Medicine published in *PLoS ONE* that when an adult cell is reprogrammed into the embryonic-like state, the slate is not wiped clean – cells still have residual gene activity of their original cell type. This suggests that for a cell to be completely reset, more steps might be needed, or certain cell types might be better candidates for reprogramming. Also in February, a study in the journal *Stem Cells* by researchers of Advanced Cell Technology, a stem cell company in Worcester, Massachusetts, found that blood vessel and retinal cells made from iPS cells aged rapidly. One thing is clear, as long as iPS cells have differences, even slight, their use will be limited, both as potential therapy and as tools to study the origins of disease or test drugs. It is appropriate to end this section the same as Section 14.3: Time will provide the answers to questions such as whether stem cells are ultimately suitable for therapeutic applications and for which disorders, but primarily with which stem cells: embryonic, adult, or induced pluripotent stem cells, or perhaps all three?

14.8. IN CONCLUSION

Stem cell research is in a critical transition phase at the moment. The first "stem cell products" have reached the clinical test phases and the market is approaching. In fact an internet search for stem cell therapies results in more than 200 companies that claim to grow stem cells, inject them back into the patient and cure almost any condition (*CRC News* 1 July 2010). Researchers from the Stanford Institute for Stem Cell Biology and Regenerative Medicine warn about these online stem cell therapies in the 2 July 2010 issue of the journal *Cell Stem Cell*. In this issue, Dr. Irving Weissman, director of this institute, warns of the potential risks to patients and describes practices and guidelines to assess the validity of internet claims, such as being wary of clinics that advertise results mainly through patient testimonials. The researchers have launched a website to educate and protect patients from unproven stem cell therapies sold online that can be dangerous and very costly. This website[103] from the International Society for Stem Cell Research (ISSCR) includes questions to ask potential clinics, and users can submit a specific website for the society to investigate. When a company or clinic is submitted for investigation, the society will evaluate whether a medical ethics committee is involved to protect the rights of patients and whether the proposed treatment will be supervised by an official regulatory body such as the US FDA. This is an excellent, very informative and sobering website, including video messages from stem cell experts. Visiting this website is a must for people considering stem cell therapy. The ISSCR has also issued key guidelines for the translation of stem cell research into the clinic. These guidelines are summarised in Textbox 14.3 and come from the review *The bioethics of stem cell research and therapy* (Hyun, 2010).

It was a very long time before Geron got permission from the FDA to start the first clinical trials with human embryonic stem cells (Section 14.2). This reflects the uncertainty that still surrounds the regulations on such clinical trials. Questions about the suitability of the regulations in question are increasingly being asked, but have until now been obscured by ethical controversies. Regulations appropriate for these times are essential to ensure adequate safety and to gain the trust of the public, without erecting excessive obstacles to the development of these products. In 2008 the EU led the way with its *Advanced Therapy Medicinal Products* regulation (von Tigerstrom, 2008). On 2 June 2010 the European Science Foundation (ESF) released their 38th Science Policy Briefing: Human Stem Cell Research and Regenerative Medicine – A European Perspective on Scientific, Ethical and Legal Issues[104]. In their press release[105] of 24 June 2010

[103] www.closerlookatstemcells.org//AM/Template.cfm?Section=Home
[104] www.esf.org/publications/science-policy-briefings
[105] www.esf.org/media-centre/press-releases.html

TEXTBOX 14.3.

Summary of key ISSCR guidelines for the translation of stem cell research into the clinic (Insoo Hyun 2010).

- *Investigators involved in preclinical or clinical research involving stem cells or their direct derivatives should act within the ISSCR guidelines and other relevant policies and regulations.*
- *Clinical research involving stem cells or their direct derivatives should be reviewed by human subject review committees supplemented with experts in stem cell science.*
- *Donors and patients need to give well-informed written consent, and they should demonstrate their understanding of the involved risks.*
- *Scientists and regulators should work to develop common reference standards.*
- *Appropriate quality standards and management systems for manufacturing cells need to be developed.*
- *Sufficient preclinical studies in relevant animal models need to be performed.*
- *Cells to be used in clinical trials must be extensively tested for potential toxicities, including tumorigenicity, in vitro and in animal studies.*
- *Patients should be monitored for long-term health effects and adverse events reported in a timely manner.*

---

concerning this briefing they summarise the stem-cell legislation in Europe:

- *Twenty-five countries have adopted legislation which explicitly prohibits human reproductive cloning (excluding Poland, Lithuania and Ireland as well as Croatia and Luxembourg).*
- *Belgium, Sweden, UK, Spain, Finland, the Czech Republic and Portugal allow human embryonic stem cell research and the derivation of new human embryonic stem cell lines from supernumerary (in excess) in vitro fertilisation embryos by law. The same countries allow somatic cell nuclear transfer by law except Finland and the Czech Republic who neither prohibit nor allow it.*
- *Belgium, Sweden and the UK have adopted legislation to allow the creation of embryos for research purposes under strict conditions*
- *Seventeen countries allow the procurement of stem cells from supernumerary embryos.*
- *Bulgaria, Croatia, Cyprus, Luxembourg, Romania and Turkey have not adopted legislation regarding human stem cell research.*

It is clear that further harmonisation can do no harm: on the contrary, it is in our opinion a must to clear the way for legal, reliable, scientifically proven stem cell therapies!

N.B. The NIH also has a very informative website on stem cells[106].

[106] stemcells.nih.gov/info/basics/basics1.asp

14.9. SOURCES

Anonymous. (2009, 26 June). N.Y. to pay for eggs for stem cell research. *The Washington Post*.

Baker, M. (2009). Stem cells: Fast and furious. *Nature, 458*(7241), 962-965.

Cyranoski, D. (2009). Mice made from induced stem cells. *Nature, 460*(7255), 560-560.

Dimos, J. T., Rodolfa, K. T., Niakan, K. K., Weisenthal, L. M., Mitsumoto, H., Chung, W., *et al.* (2008). Induced pluripotent stem cells generated from patients with ALS can be differentiated into motor neurons. *Science, 321*(5893), 1218-1221.

Glaser, V., & Hurlbut, W. B. (2005). Personal profile - An interview with William B. Hurlbut. *Rejuvenation Research, 8*(2), 110-122.

Gruen, L., & Grabel, L. (2006). Concise review: Scientific and ethical roadblocks to human embryonic stem cell therapy. *Stem Cells, 24*(10), 2162-2169.

Hurlbut, W. B. (2005). Altered nuclear transfer as a morally acceptable means for the procurement of human embryonic stem cells. *Perspectives in Biology and Medicine, 48*(2), 211-228.

Hyun, I. (2010). The bioethics of stem cell research and therapy. *Journal of Clinical Investigation, 120*(1), 71-75.

Maherali, N., Sridharan, R., Xie, W., Utikal, J., Eminli, S., Arnold, K., *et al.* (2007). Directly reprogrammed fibroblasts show global epigenetic remodeling and widespread tissue contribution. *Cell Stem Cell, 1*(1), 55-70.

Mummery, C., Van de Stolpe, A., & Roelen, B. (2007). *Stamcellen* (2nd edition ed.). Amsterdam, Veen Magazines.

Okita, K., Ichisaka, T., & Yamanaka, S. (2007). Generation of germline-competent induced pluripotent stem cells. *Nature 448*, 313-318.

Park, I. H., Arora, N., Huo, H., Maherali, N., Ahfeldt, T., Shimamura, A., *et al.* (2008). Disease-specific induced pluripotent stem cells. *Cell, 134*(5), 877-886.

Park, I. H., Zhao, R., West, J. A., Yabuuchi, A., Huo, H. G., Ince, T. A., *et al.* (2008). Reprogramming of human somatic cells to pluripotency with defined factors. *Nature, 451*(7175), 141-146.

Stadtfeld, M., Apostolou, E., Akutsu, H., Fukuda, A., Follett, P., Natesan, S., *et al.* (2010). Aberrant silencing of imprinted genes on chromosome 12qF1 in mouse induced pluripotent stem cells. *Nature, 465*(7295), 175-181.

Takahashi, K., Tanabe, K., Ohnuki, M., Narita, M., Ichisaka, T., Tomoda, K., *et al.* (2007). Induction of pluripotent stem cells from adult human fibroblasts by defined factors. *Cell, 131*(5), 861-872.

Takahashi, K., & Yamanaka, S. (2006). Induction of pluripotent stem cells from mouse embryonic and adult fibroblast cultures by defined factors. *Cell, 126*(4), 663-676.

Varmus, H. (2009). *The art and politics of science*. New York, WW Norton & Company.

von Tigerstrom, B. J. (2008). The challenges of regulating stem cell-based products. *Trends in Biotechnology, 26*(12), 653-658.

Wernig, M., Meissner, A., Foreman, R., Brambrink, T., Ku, M. C., Hochedlinger, K., *et al.* (2007). In vitro reprogramming of fibroblasts into a pluripotent ES-cell-like state. *Nature, 448*(7151), 318-324.

Yu, J., Vodyanik, M. A., Smuga-Otto, K., Antosiewicz-

Bourget, J., Frane, J. L., Tian, S., *et al.* (2007). Induced pluripotent stem cell lines derived from human somatic cells. *Science, 318*(5858), 1917-1920.

Zhao, X. Y., Li, W., Lv, Z., Liu, L., Tong, M., Hai, T., *et al.* (2009). iPS cells produce viable mice through tetraploid complementation. *Nature, 461*(7260), 86-90.

part four

Epilogue

Part 4: Epilogue

In Greek mythology Cassandra is one of the daughters of Priam, the king of Troy, who lived during the Trojan war. She was blessed by the god Apollo with the gift of prophecy. The *Cassandra syndrome* refers these days to an ominous prediction that later turns out to be true. Will the predictions of the opponents to modern biotechnology also turn out to be Cassandra prophecies? It certainly seems unlikely now. Compared to other revolutionary technologies, the calamities caused by modern biotechnology after more than 35 years are non-existent. The doom scenarios concerning modern biotechnology are very different from those of Cassandra in another way too. The more Cassandra warned people of an approaching disaster, the less they believed it would happen. In the figurative sense, Cassandra therefore stands for a prophet of doom, whom nobody believes. That can't be said of the opponents of modern biotechnology, who manage to attract attention and support from all possible media.

CASSANDRA

Part 4: Epilogue

MODERN BIOTECHNOLOGY:
FOR BETTER OR FOR WORSE?

We started this book by asking

Isn't biotechnology harmful?

We also said that by the end of the book, we would have drawn the following conclusion:

Biotechnology doesn't have to be harmful!

We have tried in this book to convince the reader of this proposition. Now that it is finished, we the authors are more convinced than ever. So we hope that, having read the book, you too can genuinely subscribe to this point of view. In the writing of this publication we have used literally thousands of articles from websites, newspapers, technical and scientific journals, books, encyclopaedias, digital newsletters, annual reports, other reports, and so forth. We have borrowed many sentences, especially from scientific journalists. We have tried to create an anthology of the many things that have been written in the area of modern biotechnology for the layman. These are frequently referred to in the text and the more scientific ones also in the list at the end of each chapter. Websites are taken up as footnotes. To facilitate visiting them, they are also available as a link on the publisher's website.[107] We have spent years writing what you have just read. However, developments in modern biotechnology move at such high speed that we have repeatedly had to rewrite parts of the text. We cannot therefore guarantee that references have not accidentally been omitted, and we offer our apologies should this be the case.

We have restricted ourselves to the subjects that have caused or are still causing controversy. We do, however, realise that there are many more interesting topics in modern biotechnology, for example, the dawning of the DNA era in forensic research. We have also made no mention of bio-nanotechnology, bioinformatics, systems biology or synthetic biology, which are closely related to or follow on naturally from modern biotechnology. All these modern biotechnologies rely on advanced scientific research and practical entrepreneurship, and their effect on society is huge.

As with all technologies, the influence of biotechnology on society can be used for good or bad. The decision lies

---

[107] www.wageningenacademic.com/modernbiotech

with the user of the technology. The biotechnology itself is no more than a means. The biotechnological revolution, which began in the early '70s of the 20th century, will undeniably greatly affect the appearance of the 21st century, for better or worse.

The scientific journalist Jan Blom expressed a similar view in the final chapter of his book 'Biotechnologie in Nederland' (Biotechnology in the Netherlands), which was published in 1985. Now, 25 years later, this effect is very noticeable. It is also clear that developments are moving even faster and their impact is even greater than was initially predicted. Therefore, it is of the utmost importance that modern biotechnology remains a prominent point of order on society's agenda, and that this continues to be food for discussion. We must all have a say in deciding what is permitted, what is not permitted! This new Pandora's box must be carefully and skilfully opened, so that we release the gifts and not the curses. We hope that this book has helped.

GM = genetically modified, also called transgenic in case of GM plants and GM animals
GMO = genetically modified organism

to everything
there is a season

TELLING TIME

JUDAH L.
MAGNES
MUSEUM

BERKELEY
California

This catalogue is published in conjunction with the exhibition TELLING TIME: *To Everything There Is a Season* held at the Judah L. Magnes Museum. May 2000–2002.

CATALOGUE CONTRIBUTORS

FRED ASTREN, *Associate Professor of Jewish Studies, San Francisco State University*

SHEILA BRAUFMAN, *Curator of Painting and Sculpture*

BILL CHAYES, *Curator of Film, Video, and Photography*

MICHAL FRIEDLANDER, *Blumenthal Curator of Judaica*

TOVA GAZIT, *Blumenthal Rare Book and Manuscript Librarian*

FLORENCE HELZEL, *Curator of Prints and Drawings*

KIM KLAUSNER, *Archivist, Western Jewish History Center*

SUSAN MORRIS, *Executive Director, Judah L. Magnes Museum*

CATALOGUE PRODUCTION

SHEILA BRAUFMAN, FLORENCE HELZEL, CARIN JACOBS, *Coordinators*

IRENE MORRIS, *Graphic Designer*

FRANCES BOWLES, *Copyeditor*

PHOTO CREDITS

All photographs are by BEN AILES except

SHARON DEVEAUX (#246); MARTIN ZEITMAN (#126)

This exhibition and catalogue reflect the wide range of Jewish observance, from the strict to very liberal interpretations of Jewish laws and customs.

Acknowledgments

TELLING TIME: *To Everything There Is a Season,* a watershed event at the Magnes Museum, represents several years of collaborative work involving the entire Museum staff. As Director of the Magnes, I am deeply appreciative to all the staff, for without their vision, dedication, and hard work, the exhibition and the catalogue would not have been possible. I am extremely grateful to the Magnes Museum Board of Trustees for their continued support and for their encouragement to mount an exhibition that dramatically demonstrates the mission of the Magnes Museum and the depth and breadth of its collections.

In particular, I wish to extend my profound appreciation to the exhibition coordinators, Sheila Braufman and Carin Jacobs. I am deeply grateful for the sustained and collaborative work of our curatorial team: Sheila Braufman, curator of Painting and Sculpture; Bill Chayes, curator of Film, Video, and Photography and exhibition designer; Michal S. Friedlander, Blumenthal curator of Judaica; Tova Gazit, librarian, Blumenthal Rare Book and Manuscript Library; Florence Helzel, curator of Prints and Drawings; Kim Klausner, archivist, Western Jewish History Center; Julie Ulmer, educator; Marni Welch, project manager; and Allyson Lazar and Corinne Whittal, assistant registrars. I express great appreciation to our development and marketing department: Shirley Gerzon, director, Carin Jacobs, marketing associate, Paula Friedman, press relations, and Genevra Tehin, assistant to the department. I would also like to thank Denise Childs, financial manager; Tamar Cohen, gift shop manager; Julia Bazar, archivist; Elizabeth Levinsky, receptionist; Herbert Singer and Jack Hyams, fine arts volunteers; and the docents and volunteers of the Museum.

I would also like to thank the following consultants: Fred Astren, associate professor of Jewish Studies, San Francisco State University; L. Thomas Frye, chief curator emeritus, Oakland Museum; Rabbi Stuart Kelman; Ruth Levitch, educator; and Roslyn Tunis, museum consultant. For the exhibition design and installation, I thank Bill Chayes; Irene Morris, graphic designer; Robin Clark, chief preparator; Travis Sommerville, Sven Atema, and Donna Bowman, preparators; Denise Fordham and Denise Krieger Migdail, mounting; Ed Tannenbaum for the entry monitor display; Julie Prosper, Denise Krieger Migdail, and Karen Zukor, conservators; Betsy Krugliak of The Pacific Group; and Rachel Trachten, Sharon Tanenbaum, Elayne Grossbard, Elizabeth Friedman and Dawn Hawk.

This exhibition incorporates perspectives on time and life cycle observances from many cultures. For these, I am grateful to Muketa Goel, chairperson of the Human Services Committee, D.V. Giri, R. Nagaraja Rao, and members of the Board of Directors of the Hindu Community and Cultural Center in Livermore; Robert L. Haynes, senior curator at the African American Museum and Library in Oakland; Bea Carrillo Hocker, specialist in Mexican art and culture; Manni Liu, executive director and curator at the Chinese Culture Center of San Francisco; Eric Moon, of American Friends Service staff and member of the Service Committee's Death Penalty Project; Vandean Philpott, director of the San Francisco African American Historical and Cultural Society; Darryl Babe Wilson, Ph.D., *Iss* (mother), *Aw'te* (father) from the north-east corner of California.

The Magnes Museum family expresses its profound gratitude to our founding director, Seymour Fromer, now Director Emeritus. It is also with deep appreciation that I thank our many donors, both of objects and of funds. It is a result of their dedication and devotion to the Museum that the mission of the Judah L. Magnes Museum is sustained and achieved.

TELLING TIME: *To Everything There Is a Season* exhibition and catalogue are made possible through generous grants from The Art and Culture Council of the Judah L. Magnes Museum, Bellarmine College Preparatory, The Jewish Community Endowment Fund of the Jewish Community Federation of San Francisco, the Peninsula, Marin, and Sonoma Counties, The Jewish Community Foundation, affiliated with the Jewish Federation of the Greater East Bay, The Living History Centre of the Endowment Fund of the Marin Community Foundation, and Northern Trust Bank. ■

Executive Director's Statement

In June 1998, as the local and world press started a countdown to 2000, the entire staff of the Judah L. Magnes Museum began to plan for a major, interdepartmental, interdisciplinary exhibition. During this initial planning phase, as various exhibition themes were explored, the subjects of science and the future seemed to recur. The world, it seemed, including the Magnes staff, was focused on a great demarcation, the impending change on the calendar from 1999 to 2000.

The subject of time is complex but universal, bridging and yet separating cultural experiences. Questions arose: What is Jewish time and how does it inform our collections? Does cultural time exist? If not, how do we account, for example, for differences in the Jewish calendar, the Chinese calendar, the Gregorian calendar, and the Muslim calendar? People from different cultures are existing in time and space at the same moment, under the same sun, stars, and moon and yet their cultural constructs may be different.

The collecting mission of the Magnes Museum, by its breadth of years and of geography, inherently relies upon an interpretation of time. As the Magnes collections reflect the complexity and diversity of the Jewish experience throughout history and throughout the world, an exploration of cultural time became compelling. Thus, an exhibition was conceived and created, though most certainly not in the biblical time of six days.

Our exhibition includes the sacred and secular, the serious and the whimsical. All are part of a cultural whole; part of the human life cycle. Our collections serve as a source for scholarship, knowledge, and participatory dialogue that, we firmly believe, will cultivate respect and understanding between community members of different backgrounds. Our mission concludes with this statement: *We promote understanding by fostering dialogue and exploring links between Jewish and other cultures.*

Our mission is timely. The time is now. Welcome to The Magnes Museum.

Susan Morris, Executive Director

ולאחר כד בלי הפסק הפסק ילמר פרשת חטאת חיל מדלאסך לבופר כנגד ה
הסכינה מרומז על דרך הקבלה על הסכינה בידרוט ליודעי חן ✦

אשת חיל מי ימצא ו
ורחוק ופנינים
מכרה ✦ בטח
בה לב בעלה ,

וטלל לא יחסר ✦ גמלתהו טוב ✦ ולא
רע כל ימי חייה ✦ דרשה לאור ופשתים
ותעש בחפץ כפיה ✦ היתה כאניות
סוחר ממרחק תביא לחמה ✦ ותקם
בעוד לילה ותתן טרף לביתה ✦ וחק
לנערותיה ✦ זממה שדה ותקחהו
ופרי כפיה נטעה כרם ✦ חגרה בעוז
מתניה ותאמץ זרועתיה ✦ טעמה כי
טוב סחרה לא יכבה בלילה נרה ✦
ידיה שלחה בכישור וכפיה תמכו פלך
כפה פרשה לעני וידיה שלחה לאביון
לא תירא לביתה משלג כי כל ביתה לב
שנים ✦ מרבדים עשתה לה שש וארגמן
לבושה ✦ נודע בשערים בעלה

242 *Sefer Olat Shabbat (The book of Shabbat burnt-offering),* 1726

The scribe and illuminator of the manuscript did not sign his name or identify the place where it was produced. Based on its similarities in script to manuscripts of the same period, we can assume that this example is of Moravian origin. Because this style of printing, known as "Amsterdam Printing," was acknowledged by the Moravian scribes, we can also conclude that this manuscript was from that region. A common practice at the time was for wealthy Jews to commission hand-illuminated manuscripts for private use or gifts. Scribes also prepared manuscripts for public use, and those they did not always sign. ■ *Olat shabbat* is a kabbalistic work, consisting of poems for *Shabbat* and explanations of the *Shabbat* rituals. It is apparent that these illuminations were modeled after printed books. As shown, the depiction of a woman and the words of praise from her husband "*Eshet hayil mi yimtsa…*" Who can find a virtuous woman? (Proverbs 31:40), is part of the *Shabbat* rituals and appears in many printed books.

JEWISH PERCEPTIONS *of* TIME:
CYCLICAL, CIRCULAR, LINEAR, AND LAYERED

by Fred Astren

Many people in modern society assume that time is a scientifically ordained constant in our lives that is not subject to interpretation or multiplicity of meaning. To the contrary, a close look at Jewish culture and circumstance reveals a complex set of approaches to time. Over the centuries Jews have expanded and eradicated time, compressed and structured time, and generated varieties of past, present, and future. Often Jews have considered time to operate cyclically, repeating and replicating itself. At the same time they have considered time to be linear, moving in one direction from a beginning toward an end.

Cyclical Time

Cyclical approaches to time in Jewish thought are evident in the calendar. One important principle of the Jewish calendar provides for linking its festivals and other holy days with the seasons. That is, Jewish tradition has demanded that particular holidays be observed at certain times of the year. This linkage originated in antiquity at a time when the ancient Israelites were virtually indistinguishable from their Canaanite neighbors, whose religion was imbued with a mythical understanding that connected the seasons, the gods, and agriculture. When worship of Yahweh, a specifically Israelite god, differentiated the Israelites from the Canaanites, many traditional cultural assumptions were preserved, thus perpetuating in Judaism the link between religion and the seasons.

Different names for the Passover holiday illustrate the creativity that blended season, livelihood, and story into calendrical time in ancient Judaism. The agricultural season is commemorated in the Hebrew, *hag ha-matzot,* the Festival of *Matzah,* when the annual grain harvest is celebrated in the spring with dietary rituals that transform the preparation and eating of bread. Because bread is wholly the product of a settled agricultural lifestyle, the springtime seasonality commemorated here speaks of Israelites settled in villages, cultivating fields, and benefiting from a peaceful secure society. Another kind of ancient Israelite is remembered in Passover's other name, *hag ha-pesach,* the Festival of Passover. The central ritual activity connected to this appellation is the Passover sacrifice that in antiquity was a lamb. Lambs are products of the flock, emblematic of a pastoral society in which ancient Israelites need not have lived as farmers in settled villages, but as seminomadic shepherds following their flocks from pasture to pasture. The origin of such a springtime sacrifice may be traced to an appeasement offering presented to the shepherds' god, as if to say: "Take this new lamb of this year's flock, but in exchange, do not take any more lambs during the coming year."

Jewish creativity does not rest on the fact that these ancient and perhaps premonotheistic practices were preserved and incorporated into Judaism, but on the fact that the biblical narrative gracefully and seamlessly folds these practices and their times of year into the story of the Exodus from Egypt. Springtime eating of unleavened bread and the lamb sacrifice are elemental to the story of the Israelites' liberation from Egypt. Distant and vague memories of life in the ancient Near East became elements of the story of God's redemption of the people. Ancient farmer and shepherd are not lost to the mists of time, but are layered into monotheism, their seasonal rituals and foods continuing to be celebrated and experienced.

The biblical narrative goes further to combine the seasons and the world of meaning by fixing the pilgrimage festivals of *Pesach, Shavuot,* and *Sukkot* in the calendar linked to the seasons and sequenced according to the Torah's narrative. That is, the springtime liberation of *Pesach* is followed in the summer by revelation at Mount Sinai commemorated on *Shavuot,* which itself is followed by commemorating the Israelites wandering in the wilderness on *Sukkot* in the fall. The temporary dwellings used to commemorate *Sukkot* evoke both the wilderness of the Torah and the guardposts found in the fields during the autumn harvest. Again, nomadic shepherds and sedentary farmers are found in a stratum just below the surface of the calendar's structure, layering the seasons, the agricultural and pastoral years, and the story of the Torah into a single composition of meaning and religious activity.

The building blocks of this elegant religious architecture are the rituals that accompany the festivals. As they experience the festivals Jews participate in the Exodus and the Sinai experience each year, thereby reifying a primary principle that God speaks through the action of history and that he is the master of time. By participating in ritual Jews actually re-enact events from the distant past, as it says in the Passover *Haggadah,* "This is done because that which God did for me when I came forth out of Egypt." In so doing, time as we normally think of it is collapsed and the distinctions between past and present are obliterated.

In fact, Jewish ritual often loses its commemorative associations and becomes a spiritual act of immediacy that sanctifies the moment, bringing the holy into the secular, the sacred into the profane. This immediacy further obliterates time as we know it to create what some have called "sacred time" or what might be referred to as "no time." When we consider Jewish ritual in this manner, we begin to approach the realm of the mystic, where human references to time, space, and self are dissolved to create an eternal "now" through which one can see the world from the perspective of God.

The power of ritual to manipulate time also functions in a unique way for Jews when the creation of ritual time simultaneously creates ritual space. Anyone who has lived or participated in *Shabbat* (the Jewish Sabbath) in an observant home knows that the quality of time and space seem to be psychologically yet tangibly transformed in astonishing ways that are inexplicable to the modern secular mind. Careful observance of religious laws limit and prohibit activities that would be considered normal during the rest of the week, such as physical labor, buying, selling, driving a car, or using an electronic device. These laws generate a difference between the Sabbath and the rest of the week. The laws further encourage activities particular to the Sabbath, such as communal worship, family celebration, study of sacred texts, and personal rest. When Jews manipulate time by observing Sabbath rules, they simultaneously create spaces in synagogues, homes, and even entire communities in which sacred time is the norm and the other days of the week seem like distant places.

Furthermore, Jews have used ritual and religious law to create spiritual and conceptual spaces in order to compensate for the loss of place and space that they suffered after the destruction of Jerusalem and the Jewish Temple by the Romans in 70 C.E. The Sabbath and other observances gave Jews the opportunity to experience space and place even though they were living in exile, outside the homeland of Israel, without political power in lands sanctified by the churches, mosques, and shrines of other dominant religions.

The cycles of time that occur in the calendar year for all Jews, all the people of Israel, are replicated in the cycles that occur in individual lives. The course of a Jewish lifetime is divided up and at the same time made whole by events that mark the passing of time and that generate social and spiritual transformations for the individual. The movement from birth rituals (naming for boys and girls and circumcision for boys) to marriage creates a cycle that completes itself when marriage yields to new birth and another ritual of naming and possibly circumcision. The circle is broken when death interrupts the cycle. Important rituals for marking that time and for confronting grief and moving toward healing are prescribed. These events and their attendant rituals cannot be understood solely as something experienced by and done for the individual. To the contrary, in these moments all the community participates. In the case of death, this organic social connectedness discourages individual mourners from being lost to personal grief and loss that is too much to bear, and encourages the community to grow and reconstitute itself from generation to generation as it celebrates and encourages family, community, and an affection for the Jewish people. The life cycles of Judaism expand time horizontally to include the entire community, becoming larger than one single life and all its joys and sorrows. And they expand time vertically to bridge the gap between generations to encompass centuries of lifetimes.

shown above
155 Circumcision bowl, detail

Linear Time

The presence and function of linear time in Judaism is grounded in the Hebrew Bible, but it challenged the rabbis of the past, and provokes and asserts itself in the modern Jewish mind. The very idea of Creation as described in the beginning of Genesis produces a linear concept of time. Jewish tradition has asserted that God created the world out of nothing, bringing into being a moment before which there was nothing and after which all existence and the procession of time follows. This idea of time as a line creates context for the Bible by providing the ground from which God speaks. By making history itself a manifestation of the divine plan, the Bible describes a God who speaks through the actions of the people of Israel and his messengers, the prophets. Even God's miracles function as direct manipulation of worldly events that proclaim his mastery and power.

This linear and historically minded biblical view is most forthright in the books of the Former Prophets, whose narratives move the reader with tales of ancient Israel's kings and warriors, God's messengers, and the world's evildoers. The centrality of God and his relationship with the people of Israel reduces historical events to symptoms of the divine plan. The Babylonians, who destroyed Jerusalem and the Temple in 587/6 B.C.E. are mere pawns in the game, and themselves summarily reduced to defeat and servitude by the Persians. Lineality in the biblical perspective of time is further reinforced by the importance of genealogy as a way to explain the people's identity, where its tribes and clans reside, and who can claim status and power in a tribal and monarchical society. Mirroring the place of the larger group in ritual life, much of the biblical story describes the linear movement of the Jewish people from the past toward the future through generations.

Centuries later the rabbis were challenged both by inheriting from the Bible this linear model of time, and by the contradiction that resulted from their own emphasis on sanctification in the present. The central theological principle of God's revelation allowed the Bible to be read less as a text from the past and more as an eternal document whose message is most significantly applied to sanctify the here and now. The application of ritual and the observance of divine command neither depended upon a specific past nor demanded any particular type of future. "There is no before or after in Torah."[1] For the enormous expenditure of rabbinic intellectual and spiritual capital the Bible provided all the past that was needed, even if it was a distant past. For that past described the origins of the Law and the operating principles for its operation.

Such a traditional view is imagined, in a recent novel, for small Eastern European Jewish communities known as *shtetls*. "Half the time the *shtetl* just wasn't there: it was in the Holy Land, and it was in the remote past or the remote future, in the company of the Patriarchs or prophets or of the Messiah. Its festivals were geared to the Palestinian climate and calendar…it prayed for the subtropical (early and late) rains indifferent to its neighbors, whose prayers had a practical, local schedule in view."[2]

In contrast, some interpreters projected the distant biblical past into the present to prophetically claim that the biblical narratives speak to contemporary history and not just to the epoch of the Israelites. Like the sect of the Dead Sea Scrolls, medieval Karaite Jews believed that biblical prophetic texts spoke to them regarding contemporary Muslims or their opponents, the rabbis, and not only about biblical priests, Assyrians, and Babylonians. The difference between past and present is obscured when historical time and specificity are eradicated in this manner.

The rabbis themselves found historical time to be of little consequence. The great medieval rabbi, Maimonides, belittled the reading of historical works. What little historical information that is found in the classical rabbinic compilations of law and lore, the Mishnah and Talmud, is there only to fulfill ideological and legal purposes of the rabbis as evidence for an interpretation of the Law, as a didactic device to support moral principles, or to centralize the people of Israel in the maelstrom of world events — all signifying that God is master of the world and of time. The Talmud centers world history around the moral behavior of the Jews: "When Solomon married Pharaoh's daughter, Gabriel descended and planted a reed in the sea, and it gathered a bank around it, on which the great city of Rome was built."[3] In this textual interpretation, or midrash, the origin of Rome, enemy and persecutor of the Jews, is the result of sin in Israel, not of historical cause and effect.

At the same time that the rabbis compressed and took apart a linear concept of time, they also embraced it in many ways. Elaborating on the biblical idea of a beginning of time, the rabbis carefully constructed an ideology for its end, which included the Messiah, the Day of Judgment, and ideas about the immortality of the soul and resurrection. Now the line could be imagined to go from one point to another. Jews could look forward to a spiritual future and God was truly imagined as the master of time. In addition, the rabbis rationalized their own authority by connecting rabbinic Judaism with the distant past of the Bible by constructing a chain of tradition that demonstrated the generation-to-generation transmission of teaching from Moses to the prophets to the rabbis themselves. Linear time proved most useful for rationalizing the rabbinic innovation of Judaism in early centuries of the common era.

The rabbis managed both cyclical and linear notions of time by careful mediating between the immediacy of sanctification and the passing of time, thereby reconciling real-life experience with knowledge of the past, divine power with human effort, hope with justice, and despair with mercy. This careful balancing act testifies to the creativity of the Jews and the complexity of their religious culture and of the societies that it engenders. It interweaves, interleaves, and encompasses: explaining the covenant relationship of God, Torah, and Israel; describing good and evil; contextualizing the Jews among the nations and in the world; and placing the individual in the group.

Rabbinic traditional approaches to time and the past were blown apart by the Jewish experience of modernity. Over the course of the past five hundred years ideas of nation, individual, and God have been radically transformed. Gradually and for the first time Jewish writers began to think historically, trying to recover a Jewish past that was lost in the experience of minority life under Christianity and Islam, or that was obscured by traditional rabbinic interpretation. Jewish participation in modern academic fields such as history, philosophy, and the sciences, and Jewish yearning for a modern national existence demanded a historical sense of time that would yield a modern identity — a Jewish community with a sense of its historical past, composed of people making individual choices about identity, and unencumbered by supernaturalism and religious obscurantism. The past became specific for modern nationalist Jews, Zionists, as they imagined a future of Jewish nationhood, or for other Jewish modernists, who looked forward to civil equality and an end to discrimination and antisemitism in Europe and elsewhere.

JEWISH WOMEN'S
CONFERENCE AND FAIR

saturday, january 28th 1978

Today, modern Jews are keenly aware of both their mythical and historical pasts and futures. They participate in sacred time with family and congregation, but carry the historical baggage of the past two hundred years during which European Jews were catapulted out of the ghetto, the social standing of Middle Eastern Jews was radically transformed, and the majority of the world's Jewish population faced murder or was transplanted to new environs. At the same time, very traditional Jews continue to embrace classical rabbinic constructs of time in resistance to the pressures of modernity. Modern Jews have learned to modulate between these poles of experience, sometimes partaking in sanctified and mythical time, and at other times participating in the inexorable progress of events and circumstance that marks the Jewish presence in the contemporary world. As Jews look forward, their depth of tradition, richness of culture, and varieties of experience will undoubtedly give birth to new approaches in the future of Jewish time. ▪

Fred Astren is Associate Professor of Jewish Studies at San Francisco State University. Dr. Astren, whose specialties are the Karaite Jews and the Jews of Islam, has written several books and articles on Jewish subjects.

---

1. From the Babylonian Talmud, the sixth-century rabbinic compilation of Jewish law and legend, Tractate Pesahim, folio 6b.
2. Conor Cruise O'Brien, *The Siege: The Saga of Israel and Zionism* (New York, 1986).
3. Babylonian Talmud, Tractate Shabbat, folio 56b.

shown above
136 Jewish Women's Conference and Fair flyer, 1978

Introduction

In today's fast-paced world, we constantly consult watches, clocks, and calendars and often schedule ourselves months in advance. Before calendars, the ancients told time by the sun, moon, and stars. The cycles of nature were critical for survival and most celebrations revolved around the seasons. Today, the calendars of many cultures reflect holidays still celebrated according to the seasons.

The Judah L. Magnes Museum presents TELLING TIME: *To Everything There Is a Season* to encourage an understanding of the similarities and differences in the ways people mark time. Structured by the seasons of the year, the exhibition highlights the interwoven cycles of nature, human life, and culture through Jewish holidays and life-cycle events, emphasizing their time-based concepts. People of various cultures share many of these concepts, remembering the past, transmitting their culture from generation to generation, and persevering against assimilation and persecution.

According to the sage Rabbi Abraham Joshua Heschel, "Judaism is a religion of time, aiming at the sanctification of time." Religious practices are designed to make Jews aware of every sacred moment. The rituals for the Sabbath (*Shabbat*) for instance, are meant to create a sense of sacred time to be spent in rest and spiritual renewal, beginning with candle lighting before sundown on Friday and ending after sunset on Saturday, when the celebration of *Havdalah* (separation) ushers in the new week.

In the exhibition the *Shabbat* installation appears with those of other cultures that also demonstrate the demarcation of sacred time. Perspectives written by scholars of diverse cultures to emphasize shared concepts are also inserted throughout the exhibition. For instance, we consider how both Jewish mourning rituals and Day of the Dead practices in Latino tradition help us pay tribute to ancestors.

Although all creatures exist in time and are changed by it, human beings are the only ones who can manipulate time. By employing imagination and foresight along with memory, we can recapture the past and conceive of the future. ■

All works listed are from the Judah L. Magnes Museum's collections unless otherwise noted. Dimensions are in inches; height precedes width (precedes depth where appropriate). An asterisk {∗} indicates works illustrated in this catalogue. Checklists follow each section.

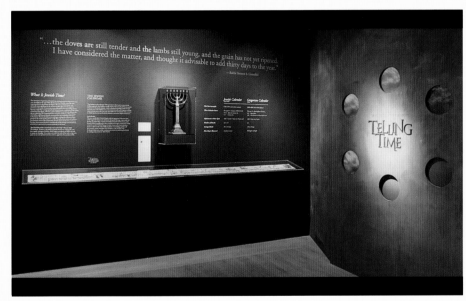

INTRODUCTION

1 Torah binder (*wimpel*)
Maker: Koppel Heller
Munich, Germany, 1815
Linen, embroidered with silk thread; and silk lining
28 1/8 x 20 1/8
Gift of Mr. and Mrs. Theodore Lilienthal
80.83

2 Calendar
Scribe: Mordecai, son of Baruch Shabtai
Calcutta, India, 1910
Ink, paint, and graphite on paper
2000.0.1

3 Seven-branched candelabrum (*menorah*)
Damascus, Syria, ca. 1924–25
Copper alloy, silver; base: wood
Gift of Mrs. Bonnie I. Henning
2000.5.1

4 *Yesod Olam*
(The Foundation of the World)
Author: Isaac Israeli, 14th century
Berlin, 1777
In Hebrew
8 1/4 x 6 3/4
Rare Book Collection
Blumenthal Rare Book and Manuscript Library

5 *Moon Ritual,* 1928
Peter Krasnow
(Russia 1887–1979 U.S.A.)
Lithograph
19 3/4 x 15 1/2
Gift of Leo Rabinowitz
74.27

11 *And There Was Light* (from the portfolio *Genesis*), 1925, Abel Pann

Pann illustrates the biblical version of the origin of the world: "And God said: Let there be light and there was light. And God saw the light, that it was good; and God separated the light from the darkness. And God called the light Day, and the darkness God called Night. And there was evening and there was morning, a first day." Genesis I:3–5. Pann's elegantly composed illustrations of the Bible were the first color lithographs produced in Israel.

AUTUMN

HEBREW MONTHS: *Tishri, Heshvan, Kislev*
GREGORIAN MONTHS: *September, October, November*

Time for the harvest. The mature crops are gathered, distributed, and stored in preparation for winter.

The Jewish year begins in the fall month of Tishri. Considered the holiest time of the year, *Rosh ha-Shanah,* the Jewish New Year, is the time of the creation of the world. *Rosh ha-Shanah* begins a time of self-examination, reflection, and repentance that concludes ten days later on *Yom Kippur* (the Day of Atonement). *Yom Kippur* is a day of abstinence and majestic liturgy. Sins are confessed and it is hoped that one will become reconciled with God and others through repentance, prayer, and *mitzvot* (good works). One view of many in Judaism about the final goal and destiny of humankind is that these good works are an integral part of the process of *Tikkun Olam* (healing or repair of the world), which will lead to the Messianic Era, a righteous society in which all individuals may maintain their dignity and freedom, and there will be peaceful coexistence of all peoples, freedom from fear and want, and cessation of war.

With the final blowing of the *Shofar* (ram's horn) that concludes *Yom Kippur,* the Jewish High Holy Days draw to a close and the focus of the Jewish community shifts from the solemnity of *Yom Kippur* to the jubilant harvest celebration of the festival of *Sukkot.* This festival is named for the huts (*sukkot*) in which Moses and the ancient Hebrews lived during their forty-year journey to the land of Canaan, the Promised Land, today known as Israel. The ninth day of *Sukkot* is known as *Simchat Torah.* This festival marks the completion of the annual cycle of reading of the Torah scroll and the immediate commencement of a new cycle, a return to the beginning of the same text, with a reading of the first verses of Genesis. The cycle may be seen to symbolize the infinite nature of spiritual progression, which always begins anew. ∎

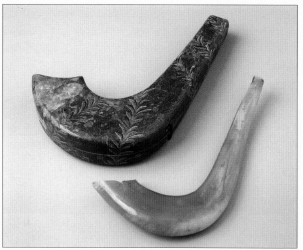

17 *Shofar* (ram's horn)
with case, 1784

16 *Shofar,*
1940, Imre Amos

53 *The Black Fire, The White Fire, The Red Fire, The Green Bring the Gold,* 1986, Edith Altman

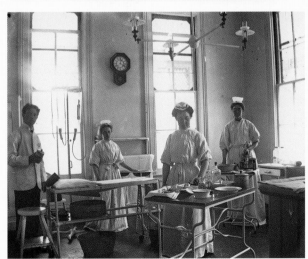

39 Photograph of nurses, ca. 1900

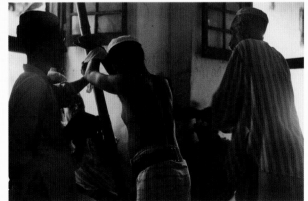

48 Ritual scourging on day before *Yom Kippur,* 1937

55 Untitled
(harvest scene), mid-1920s
Maurycy Minkowski

35 *Share,* ca. 1915
Burke (Johnstone Studios)

56 Family in *sukkah,*
1906

68 *Simchat Torah* flag
with New Year's
greeting, 1898

AUTUMN

Nature

6 *Grape Pickers,* 1941
Boris Deutsch
(Lithuania 1892–1978 U.S.A.)
Graphite and charcoal on paper
7 3/4 x 18
Gift of the Boris Deutsch estate
81.64.42

7 Girl holding grapes,
Israel, 1954
Promotional photo by
United Jewish Appeal
Gelatin silver print
9 1/2 x 7 1/2
72.0.2.55

8 Cellars of
Rishon Le Zion, Israel
Postcard
3 1/2 x 5 1/2
Gift of Walter J. Fischel
98.0.0657

9 In the fields,
Rishon Le Zion, Israel
Postcard
3 1/2 x 5 1/2
Gift of Walter J. Fischel
98.0.0658

Creation: One Perspective

10 *Creation,* 2000
Ed Tannenbaum
(b. 1953 U.S.A.)
Interactive computer animation

11* *And There Was Light*
(from the portfolio *Genesis*),
1925
Abel Pann
(1883–1963 Latvia)
Color lithograph
17 5/8 x 12 3/8
Gift of Dr. and Mrs. Leon Kolb
75.132

12 *Genesis (Bereshit)*
Translator: Robert Alter
Illustrator: Michael Mazur
(b. 1935)
San Francisco: Arion Press,
1996
"Limited to 200 numbered
copies…each copy is signed
by the artist and the translator."
Handmade paper
In Hebrew and English
17 x 12
Gift of Barry Traub
Rare Book Collection
Blumenthal Rare Book
and Manuscript Library

Rosh ha-Shanah: A New Year Begins

13 *Rabbi Reading,* 1919/20
Max Weber
(Poland 1881–1961 U.S.A.)
Woodcut with New Year's greeting
4 1/2 x 6 1/2
Museum purchase with funds
provided by the Helzel
Family Foundation
2000.12

14 *Das Yahr Des Jüden*
(The Jewish Year)
Author: Arno Nadel (1878–1948)
Illustrator: Joseph Budko
(1888–1940 Poland)
Berlin: *Verlag für Yüdische
Kunst und Kultur,* 1920
In German
9 1/2 x 11 1/2
Illustrated Book Collection
Blumenthal Rare Book
and Manuscript Library

15 Untitled (Rabbi in synagogue)
Isidor Kaufmann
(Hungary 1853–1921 Austria)
Oil on board, unfinished
14 1/4 x 10 5/8
Museum purchase with funds
provided by Dr. Elliot Zaleznik
80.48

16* *Shofar* (from the portfolio
Amos Imre 14 eredeti metszete),
1940
Imre Amos (1908–1945 Hungary)
Linocut
13 x 11
Gift of Ellen Miller in honor of
Leslie Kessler's birthday
87.30.5

17* *Shofar* (ram's horn)
with case
Beerfelden, Germany, 1784
Case: wood, gouache,
velvet, white metal
shofar:
6 1/4 x 11 1/2 x 1 3/4
case:
8 3/4 x 12 1/4 x 2 1/4
Gift of Mr. and Mrs. Irving Jonas
73.27 a & b

18 *Akedat Yizchak*
(The Binding of Isaac)
Artist: Moshe Mizrachi
Jerusalem, Palestine, 1920s
Paper, printed
22 x 16 1/2
Judaica collection
2000.0.10

19 New Year's postcard
Printed in Germany,
early 20th century
Paper, printed
5 1/2 x 3 1/2
Gift of Fanny Fox
72.42.1

20 New Year's postcard
Tashlikh
Williamsburg Art Co., New York
Printed in Germany, ca. 1915
Paper, printed
5 1/2 x 3 1/2
Gift of Solomon L. Gluck
73.43.56

21 New Year's greeting
Vienna, Austria, 1870
Ink on paper, printed
and embossed
11 x 8 3/4
Museum purchase
94.28.2

22 New Year's postcard
Publisher: Williamsburg Art Co.,
New York
Printed in Saxony, ca. 1920
Paper, printed
3 1/2 x 5 1/2
Gift of Solomon L. Gluck
73.43.22

23 New Year's greeting
Issued by: American Hebrew
Congregation "Tiferet
Yerushalayim"
Jerusalem, 1899
Ink on paper; printed image size:
11 x 9
Gift of the California Historical
Society
68.93

24 Calendar
Great Palestine orphan asylum
Diskin
Jerusalem, Israel, 5710
(1949/50)
Paper, printed
4 x 2 3/4
Judaica collection

25 New Year's card, Russia,
1908/09
Photographer unknown
Gelatin silver print
6 x 3 7/8
Gift of Aaron Rashkin
89.67.1

26 New Year's card, B. Stern
family
Postcard
3 1/2 x 5 1/2
Gift of Stuart and Beverly
Denenberg
84.35.6

27 New Year's card, Cyprus,
1948
Photographer unknown
Gelatin silver print
2 5/8 x 3 5/8
Gift of Morris Laub
84.42.2.34

28 New Year's card, Cyprus,
1949
Photographer unknown
Gelatin silver print
2 1/2 x 7 1/8
Gift of Morris Laub
84.42.2.35

29 *The Meditator,* 1945
Aaron Goodelman
(Russia 1890–1978 U.S.A.)
Granite
21 1/2 x 9 3/4 x 16
Gift of the artist
77.78

30 *Buddha in the Golan
Heights,* 1996
Judy Moore-Kraichnan
(b. 1941 U.S.A.)
Multitoned gelatin silver print
16 x 16
Gift of the photographer

*Tikkun Olam:
Toward a Righteous
Society*

31 Tray
U.S.A., early 20th century
Silver
3/4 x 20 x 15 1/8
Gift of Julius Kahn III
79.77a

32 Photograph,
Florence Prag Kahn, 1926
8 x 10
Florence Prag and Julius Kahn
Collection
Western Jewish History Center

33 Ouroboros statuette
Aris Demetrios
Bronze and marble
On loan from the Goldman
Environmental Prize

34 *Profiles of Environmental
Heroes from Around the World*
Produced by the Goldman
Environmental Prize
Producers: Robert Roll and
relatives
Beta Sp Video

35* *Share,* ca. 1915
Burke (Johnstone Studios)
Color lithograph
Published by the Jewish Relief
Campaign
40 1/8 x 30 1/4
Museum purchase
75.224

36 Medical diploma from the
University of Padua
Aloysius Foppa, illuminator
Ancona, Italy, 1682
Ink, gold and silver paint on
vellum
9 1/2 x 14 (open)
Binding: leather, tooled, gilt
Gift of Mr. and Mrs. Frederick
Weiss in honor of Mrs. Leon
Mandelson
73.8

37 Minute Book, 1887
18 x 14
Mt. Zion Hospital Collection
Western Jewish History Center

38 Executive Board Minutes
with folder, 1913
2 pages,
each 8 1/2 x 13
folder: 10 x 4 1/2
Mt. Zion Hospital Collection
Western Jewish History Center

39* Photograph of nurses,
ca. 1900
8 x 10
Mt. Zion Hospital Collection
Western Jewish History Center

40 Postcard
Donations on Erev Yom Kippur
(the eve of the Day of Atonement)
Publisher: Williamsburg Art Co.,
New York
Printed in Germany, ca. 1910
Paper, printed
5 1/2 x 3 3/8
Gift of Solomon L. Gluck
73.43.70

41 Charity (*tzedakah*) box
Turkey, inscription date, 1879
Silver
4¼ x 4 x 3
Museum purchase with funds
provided by Dr. Elliot Zaleznik
83.56.2 a & b

42 Charity (*tzedakah*) box, 1999
Convertor
Bruce Cannon, artist
(b. 1960 U.S.A.)
Steel box, bill validator,
computer display
9⅞ x 16⅞ x 9
Gift of Richard Goldman
2000.17 a & b

43 Certificate of wardenship
for *tzedakah* boxes
Issued to: Abraham Marcus,
Spokane, Washington
Printed in Jerusalem, Palestine,
date of issue: 1940
Ink and paint on paper and
gold paper seal
Image size:
14¼ x 19
Judaica collection
2000.0.12

Yom Kippur: Repentance and Reconciliation

44 Form of Prayers, for the
Day of Atonement
New York: L.H. Frank, 1854
In Hebrew and English
8 x 5
Rare Book Collection
Blumenthal Rare Book
and Manuscript Library

45 Lamp for *Yom Kippur*
(the Day of Atonement)
Cochin, India, inscription date,
1670
Copper alloy
17 x 13½
base diameter: 7¼
Gift of the Jewish community of
Parur, India
77.343 a & b

46 Gown (*kittel*) with belt and hat
Germany, late 19th–early
20th century
Linen
kittel: length 44½
belt: length 67
Gift of Mrs. Paula Pulver
75.183.11

47 *Yom Kippur* ritual,
brushing ground,
Cochin, India, 1937
D.G. Mandelbaum
(b. U.S.A.)
Gelatin silver print
2⅜ x 3⅝
Gift of the photographer
97.0.0125

48* Ritual scourging on day
before *Yom Kippur*, Cochin,
India, 1937
D.G. Mandelbaum
(b. U.S.A.)
Gelatin silver print
8 x 10
Gift of the photographer
97.0.0104-5.6

49 Flyer
On front: *Horaire des Prières
de Yom Kippour* (Times of
prayer for *Yom Kippur*)
Temple d'Ismailieh
Egypt, 1946
Paper, printed
7 x 4½
2000.0.11

Messianic Era

50 New Year's postcard
*And it shall come to pass
in the last days*
Publisher: Williamsburg Post
Card Co., New York
Printed in Germany
Paper, printed
3½ x 5½
Gift of Solomon L. Gluck
73.43.15

51 New Year's postcard
*The Messiah has come, the
Messiah is already here …*
Publisher: S. Resnik, Warsaw
and New York
Printed in Germany, early
20th century
Paper, printed
3½ x 5½
Gift of Solomon L. Gluck
73.43.78

52 New Year's postcard
Printed in Austria, ca. 1910
Paper, printed
5½ x 3½
Gift of Nell Mendelsohn
92.34.17

53* *The Black Fire,
The White Fire, The Red Fire,
The Green Bring the Gold,*
1986
Edith Altman
(b. 1930 Germany;
lives in Illinois)
Acrylic paint and oil stick
on Masonite
45 x 60
Gift of Mrs. Leon Fieldman
94.46.a-i

Sukkot: Shelter for the Journey to the Promised Land

54 Ethiopian women
harvesting, 1984
Marriam Cramer Ring
(b. U.S.A.)
Gelatin silver print
14 x 11
Gift of the photographer
93.9.14

55* Untitled (harvest scene),
mid-1920s
Maurycy Minkowski
(1881–1930 Poland)
Oil on wood
29 x 23¼
Gift of the Greta R. Windmiller
Trust
98.2.1

56* Family in *sukkah*, Wilkes
Barre, Pennsylvania, 1906
Photographer unknown
Gelatin silver print
12 x 14
Gift of Stuart and Beverly
Denenberg
84.35.5

57 *Sukkah* at synagogue
in Ernakulam, India, 1937
D. G. Mandelbaum
(b. U.S.A.)
Gelatin silver print
2 3/4 x 2 5/8
Gift of the photographer
97.0.0133

58 *Sukkot* (from the portfolio
Jewish Holidays), 1977
Pinchas Shaar
(b. 1923 Poland)
Screenprint, XXXVI/CXXV
30 x 22
Gift of M. Magidson
and Associates
79.33.5

59 *Etrog* box
Europe, inscription date,
1849/50
Base metal, paint, fabric,
wool and metal thread
4 x 6 1/2 x 4
76.272.2

60 New Year's postcard
In an etrog garden
Publisher: Williamsburg Post
Card Co.,
New York, early 20th century
Printed in Germany
Paper, printed
5 1/2 x 3 1/2
73.43.96

61 New Year's greeting card
Printed in Germany,
early 20th century
Paper, printed
4 1/2 x 4 (closed)
73.35.3

62 *Kinder Kalendar des
Jüdischen Frauen Bundes*,
1934/35
Germany, 1934
Paper, printed
8 x 5 1/2 (closed)
Judaica collection
75.33.1

Simchat Torah:
The Cycle Ends and
Begins Again

63 *Simchat Torah* (from the
portfolio *The Jewish Holidays*),
1968
Chaim Gross
(Austria 1904–1991 U.S.A.)
Color lithograph, 191/200
19 x 25
Gift of Renee and Chaim Gross
in honor of Tom Freudenheim
90.3.2.4

64 Torah shield with *parashah*
plaque for *Simchat Torah*
Maker: Franz Lorenz Turinsky
(1789–1828)
Vienna, Austria, 1807–1809
Silver, parti-gilt
18 x 12
Museum purchase
65.0.2

65 *Simchat Torah* vest
Cochin, India, post-1917
Velvet, cotton, beads, coins,
metal ornaments and silver
chain
19 1/2 x 19
Gift of the Cochin Jewish
community
75.183.34

66 New Year's postcard
Hakkafot
Publisher: Williamsburg
Post Card Co., New York
Printed in Germany, ca. 1915
Paper, printed
5 1/2 x 3 1/2
Gift of Solomon L. Gluck
73.43.72

67 New Year's postcard
Children holding
Simchat Torah flags
Publisher: Williamsburg Post
Card Co., New York
Printed in Germany, ca. 1915
Paper, printed
3 1/2 x 5 1/2
Gift of Solomon L. Gluck
73.43.65

68* *Simchat Torah* flag
with New Year's greetings
Publisher: J. Katzenelenbogen
New York, 1898
Paper, printed
9 1/2 x 12
Gift of Seymour Fromer
2000.0.13

69 Rohekar Hebrew
English calendar
Prepared by: D. M. Rohekar
Publisher: S. David
Printer: Vikas Printers
Bombay, India, 1966/67
Paper, printed, base metal
and cotton cord
14 3/4 x 9 5/8
Judaica collection

70 *Simchat Torah,*
São Paulo, Brazil, 1937
Photographer unknown
Gelatin silver print
10 1/2 x 12 5/8
Gift of Corinne Fischel-Zeffren
86.18.1

71 *Simchat Torah* procession,
Ernakulam, India, 1937
D. G. Mandelbaum (b. U.S.A.)
Gelatin silver print
10 x 8
Gift of the photographer
97.0.079-80

72 Goat skin being prepared
for parchment for *Simchat
Torah,* Ernakulam, India, 1937
D. G. Mandelbaum (b. U.S.A.)
Gelatin silver print
2 5/8 x 2 1/2
Gift of the photographer
97.0.0130

75 *Snowstorm, Jerusalem,* 1982, Neil Folberg

WINTER

HEBREW MONTHS: *Kislev, Tevet, Shevat, Adar I (Adar II in a leap year)*
GREGORIAN MONTHS: *December, January, February, March*

Winter is traditionally considered a time of dormancy, yet we may also regard it as a necessary prelude to regeneration and renewal — part of an unbreakable cycle in nature and metaphorically, in the lives of humans and cultures as well. The ground lies fallow while the rains nourish it for future growth. As spring approaches, in modern times, Jews help revive the land by planting trees and involving themselves in other environmental activities during *Tu bi-Shevat* (also called "New Year of the Trees").

In the winter of their lives people age and face death, yet it may be said that their lives are regenerated as they pass on their values and traditions, becoming links in a living chain between past and future generations. When people die, great emphasis is placed on keeping their memories alive through Jewish mourning rituals and traditions. These include *Shivah,* the seven-day mourning period; reciting the mourners' *Kaddish,* a prayer sanctifying God, at services for thirty days for a close relative and eleven months for a parent; commemorating the *yahrzeit,* or yearly anniversary of a death, by lighting a twenty-four-hour candle and reciting the mourners' *Kaddish;* erecting tombstones; and donating objects or funds in memory of the deceased.

Regeneration also applies to minority cultures as they struggle to maintain their identities over time while living in the wider society, persevering against assimilation from within and persecution from without. The story of *Chanukkah* exemplifies the challenges faced by Jews of the past when they resisted assimilation under Hellenism. *Purim* recalls a period when Jews might have been eliminated as a minority, but were saved by the intervention of Queen Esther of Persia. One of the ways that Jews have maintained their cultural identity while being involved in diverse commmunities is derived, however, less from heroism than from the tradition of interpretation, which examines issues in the context of their time and place according to Jewish law and tradition. ■

76 *Landscape*, ca. 1900
Lesser Ury

77 *Our Holidays, The Season of Rejoicing (Tu bi-Shevat)*, 1924/25
Zeev Raban

97 *Sefer ha-Hayim,* 1889

85 *Grandmother II,* 1928
Hermann Struck

98 *Kaddish,* 1997
Moshe Gershuni,
Text bt Allen Ginsberg

121 *Purim,* Cyprus camps, 1948

120 Esther scroll (*Megillath Esther*), detail, 18th century

114 *Chanukkah* wall plaque, ca. 18th century

124 *Lavater and Lessing Visit Moses Mendelssohn,* 1856
Moritz Daniel Oppenheim

127 Concordia Club
wrestling medal, 1889

129 Jewish Folk Chorus
of Petaluma poster, 1966

WINTER

Nature

73 Israel picture calendar, 1968/69
Japeth and Levin Epstein presses
Tel Aviv, Israel, 1968
Paper, printed; and base metal
9 x 7¹/₄
Judaica collection

74 *Frankfurt am Main,*
ca. 1927–33
Jacob Nussbaum
(Germany 1873–1936 Palestine)
Oil on canvas
23³/₄ x 31³/₄
Gift of Hans Wolff
86.34.1

75* *Snowstorm, Jerusalem,*
1982
Neil Folberg
(b. 1950 U.S.A.)
Gelatin silver print 11/50
14 x 14
Gift of David Backman
99.4.1.1

Tu Bi-Shevat: New Year of the Trees

76* *Landscape,* ca. 1900
Lesser Ury
(1861–1931 Germany)
Pastel on cardboard
18¹/₄ x 13³/₄
Gift of Julian Stanford
79.27

77* *Our Holidays,*
The Season of Rejoicing
Illustrator: Zeev Raban
(1890–1970 Poland)
New York: Dr. Isaac Miller,
1924/25
8¹/₄ x 8¹/₄
Illustrated Book Collection
Blumenthal Rare Book
and Manuscript Library

78 *Prayer for the Redwoods*
Author: Michael Lavigne
Designer: Lynda Fiesel
San Francisco: Temple
Emanu-El, 1997
Paper, printed
7 x 5
Anonymous loan

79 Pamphlet
Birkat Ha-Ilanot
(Blessing of the Trees)
Tunisia: Libraire Hebraique
Moderne, Mardochée Uzan
& Frère, early 20th century
Paper, printed
6¹/₄ x 4¹/₄
Blumenthal Rare Book and
Manuscript Library

80 Greeting card
Publisher: Rabbi Rosenberg
cards, Larry Chusid Creations
Portland, Oregon, U.S.A.,
ca. 1980
Paper, printed
6³/₈ x 4⁵/₈ (closed)
Collection of
Rabbi Mark Hurvitz

81 *Chanukkah* lamp
Europe, ca. 1945–1952
Ceramic
4¹/₂ x 10 x 6
Gift of Rabbi William Z. Dalin
79.70.1

The Elderly: Human Link from Past to Future

82 Ethiopian Jewish woman
of the village of Attege, 1983
Marriam Cramer Ring
(b. U.S.A.)
Gelatin silver print
14 x 11
Gift of the photographer
93.9.31

83 Old Age Home,
Palestine, ca. 1920s
Gelatin silver print
6¹/₈ x 8
Museum purchase
80.77.3.14

84 Old Age Home,
Palestine, ca. 1920s
Gelatin silver print
8¹/₄ x 6¹/₈
Museum purchase
80.77.3.15

85* *Grandmother II,* 1928
Hermann Struck
(1876–1944 Germany)
Etching
10¹/₂ x 15¹/₈
Gift of Mark Hurvitz
91.59.25

86 Pendant containing
photograph of elderly man
Eastern Europe,
late 19th– early 20th century
Silver-plate, colored glass
and silver gelatin print;
pendant:
1⁵/₈ x 1¹/₈ (closed)
photograph:
1¹/₄ x ³/₄
67.208

87 Jonathan Eybeschütz
Baltzer after Kleinhardt
From: *"Abbildungen Böhmischer*
und Mährischer Gelehrten und
Künstler"
Prague, 1773
Paper, printed
7 x 4⁵/₈
Museum purchase
Strauss Collection
76.14

88 Portraits of
Chana Yenta bat Itzhak and
Joseph Bereshtechko
New York, 1920s
Pastel on paper, muslin backing
18 x 14
Gift of Dr. Ralph Bennet
Judaica collection

89 A Genealogical Tree
of Noah's Descendants From
Whom Are Derived All the
Nations of the Earth Since the
Deluge, Genesis X
From: *"An Historical, Critical,
Genealogical, Chronological,
Etymological Dictionary of the
Holy Bible"*
Author: Augustine Calmet
(1672–1757)
London, England, 1732
Paper, printed
16 x 20
Gift of Seymour Fromer
Judaica collection
2000.0.14.1

90 Hebrew Ethical Wills
Tsavaot Geone Yisrael
Editor: Israel Abrahams
(1858–1924)
Philadelphia: Jewish Publication
Society of America, 1926
In Hebrew and English
6³/₄ x 4¹/₂
Gift of Jack and Susan Riptsteen
in memory of Rabbi Baruch
and Glady Braunstein
Rare Book Collection
Blumenthal Rare Book
and Manuscript Library

91 Greenebaum–Gerstle
Genealogy, 1954
24 x 136
Richard Sloss Collection
Western Jewish History Center

92 Photographs, Livingston
Family, 1912 and 1935
6 x 9¹/₂ and 9¹/₂ x 7¹/₂
Livingston Collection
Western Jewish History Center

93 Photographs,
Louis and Sarah Sloss
Mounted on scrapbook page,
10¹/₂ x 15³/₄
Sophie Gerstle Lilienthal
Collection
Western Jewish History Center

94 Pair of mugs
U.S.A., 1980s
Inscribed: "World's Greatest
Bubba" (Grandma) and
"World's Greatest *Zeyda*"
(Grandpa)
Ceramic
3³/₄ x 3³/₈;
diameter: 3¹/₈
Collection of
Rabbi Mark Hurvitz

*Death: Remembering
through Mourning
and Burial Rituals
and Traditions*

95 *Jüdisches Ceremoniel*
(Jewish Ceremonies);
plate no. 14
Author: Paul Christian Kirchner
Illustrator: Johann George
Pushner
Nuremberg, Germany:
Peter Conrad Monath, 1724
In German
8¹/₄ x 6³/₄
Rare Book Collection
Blumenthal Rare Book
and Manuscript Library

96 Union Home Prayer Book
Macon, Georgia: Central
Conference of American
Rabbis, 1951
Rare Book Collection
Blumenthal Rare Book
and Manuscript Library

97* *Sefer Hayim* (Book of Life)
Probably Jerusalem, 1889
Ink on parchment; cover:
olivewood
In Hebrew
10¹/₄ x 8¹/₂
Museum purchase
Rare Book Collection
Blumenthal Rare Book
and Manuscript Library

98* *Kaddish* (portfolio), 1997
Moshe Gershuni (b. 1936 Israel)
Twenty-four screenprints with
metal foil leaves, 7/48
19⁵/₈ x 25³/₄
Text by Allen Ginsberg
(1926–1997)
Museum purchase with funds
provided by the Helzel
Family Foundation in memory
of Ida Frank
98.25.1-27

99 *Yahrzeit* calendar
In memory of Herman (Hirsch)
and Golde Seidenberg
Publisher: Leopold Lengyel
Budapest, ca. 1907
Ink, graphite and gold paint
on paper and card
15 x 20
Gift of Louis Shawl
81.38.2

100 *Yahrzeit* list
Mikulas, Czechoslovakia,
late 19th–early 20th century
Pen and ink on paper
22⁵/₈ x 16³/₈
2000.0.15

101 *Life Cycle: Death*, 1996
Ezequiel Rotstain
(b. 1957 Peru; lives in Florida)
Acrylic and mixed media on
wood
39 x 15 x 9¹/₄
Gift of the artist
97.18.4

102 Electric *yahrzeit* lamp
U.S.A., 1940s
7; diameter: 4 without cord
Gift of Stephen
and Lynne Kinsey
99.9 a-c

103 Memorial lamp
U.S.A., 1980s
Inscription: Mount Sinai
Memorial Parlor and Mortuary
Glass; 9;
diameter: 3³/₈
Collection of
Rabbi Mark Hurvitz

104 Groman Mortuaries
memorial lamp box
Los Angeles, California,
U.S.A., 1980s
Paper, printed
3¹/₂ x 3¹/₄ x 9¹/₂
Collection of
Rabbi Mark Hurvitz

105 Glasband-Willen
Mortuaries *yahrzeit*
reminder card
For Fannie Hurvitz
Los Angeles, California,
U.S.A., 1980
Paper, printed
3¹/₂ x 5¹/₂
Collection of Rabbi Mark
Hurvitz

106 Memorial candle
Distributor: I. Rokeach &
Sons, Inc., Farmingdale,
New Jersey
Made in Israel, 1980s
Base metal, wax, cotton wick
and paper; 2³/₈;
diameter: 2¹/₈
Collection of
Rabbi Mark Hurvitz

107 Memorial candle
Manufacturer: "Ner Zion"
candle factory
Rivat Gat, Israel, 1980s
Base metal, wax, cotton wick
and paper; 2³/₈;
diameter: 2¹/₈
Collection of
Rabbi Mark Hurvitz

108 Print
Die Gräber der Vorfahren
(Graves of the Ancestors)
Ludwig Wilhelm Riefstahl
(Neustrelitz 1827–1888 Munich)
Berlin, Germany, 1864 or 1867
Paper, printed
13¹/₄ x 10¹/₄
Gift from the Harry B.
and Branka J. Sondheim
Judaica collection
Recent acquisition

109 Israel Picture Calendar,
1960/61
Publisher: Lion the Printer
Tel Aviv, Israel 1960
Paper, printed and base metal
10 x 6¹/₄
Judaica collection

110 Memorial plaque with
hanging light
Tunisia, inscription date, 1929
Inscribed in French:
"Simon Parienti died
August 23, 1929";
Inscribed in Hebrew:
"In memory of Simon Parienti"
Wood, silver, glass and
copper alloy
19³/₄ x 13³/₄ x 5⁷/₈
(without hanging light)
Museum purchase with funds
provided by Barbara and
Sheldon Rothblatt in memory of
Gittel and Morris Rothblatt
90.18 a-c

111 Memorial lamp
Morocco, early 20th century
Inscribed in Hebrew:
"In memory of the [honored]
woman Sara Rachel Ze'evi
may her soul be bound up in
the bond of eternal life"
Copper alloy
8³/₄ x 5³/₄ x 5¹/₂
Museum purchase
71.27

112 Photograph, Placerville
Jewish Cemetery, Abraham
Simon, 1983
Ira Nowinski
(b. 1942 U.S.A.)
14 x 11
Commission for the Preservation
of Pioneer Jewish Cemeteries
Collection
Western Jewish History Center

113 Textile fragment with
dedicatory inscription
Calcutta, India, inscription
date 1924
Cotton, velvet: embroidered with
metal and cotton threads
11¹/₄ x 15¹/₂
75.183.100

Chanukkah

114* *Chanukkah* wall plaque
Germany, probably 18th century
Gouache on paper
23¹/₂ x 17¹/₂
Museum purchase
Strauss Collection
67.1.6.15

115 *Chanukkah* lamp
Italy, 17th and 19th centuries
Copper alloy
14 x 10¹/₂
Museum purchase
Strauss Collection
61.1.4.47

116 *Feast of Lights*
Irving Amen
(b. 1918 U.S.A.)
Woodcut, A/P
22 x 16³/₈
Gift of the artist
94.35.9

117 *Chanukkah* party, ca. 1950
Photographer unknown
Kodachrome stereo
transparencies
1⁵/₈ x 4
On loan from the collection
of Denise Urdang

Purim

118 Poster, 1906
11 x 14
Ephemera collection
Western Jewish History Center

119 *Purim* plaque
Europe, 18th century
Gouache, ink, and gold paint
on parchment
11 x 8¹/₂
Museum purchase
Strauss Collection
67.1.6.17

120* Esther scroll
Megillath Esther
Holland, 18th century
Scroll: ink on parchment,
printer's ink; staves: wood
staves: height 11
scroll: height 7³/₄
Museum purchase
Strauss Collection
67.1.11.5

121* *Purim,*
Cyprus camps, 1948
Photographer unknown
Gelatin silver print
$5^1/8$ x $3^3/8$
Gift of Morris Laub
84.42.2.36

122 Making *Hamantashen* for
Purim, Calcutta, India, 1945
Major David Ball
Gelatin silver print
15 x 7
92.1.76

123 *Esther* (from the
portfolio *Proverbs*), 1979
Nikos Stavroulakis
(b. 1932 Greece)
Woodcut, 27/500
$19^5/8$ x $13^7/8$
Gift of the artist
79.50.7

Renewal: Maintaining Cultural Identity over Time

Involvement in the Greater Society

124* *Lavater and Lessing Visit
Moses Mendelssohn,* 1856
Moritz Daniel Oppenheim
(1800–1882 Germany)
Oil on canvas
28 x $23^1/2$
Gift of Vernon Stroud, Gerda
Mathan, Ilsa Feiger, Eva Linker,
and Irwin Straus in memory
of Frederick and Edith Straus
75.18

125 *Portrait of Isaac and
Jette Levison,* 1857
J. Eberhart
Oil on canvas
13 x $14^7/8$
Purchased with funds donated
by Dr. Daniel K. and Fritzi
Oxman
76.2

126* *Challah* cover
U.S.A., inscription date 1851/52
Silk brocade: appliquéd with
velvet; embroidered with
metallic thread; metallic braid;
silk and metallic tassels
$22^1/2$ x $20^3/4$ (including tassels)
Museum purchase
77.341

127* Concordia Club
wrestling medal
Maker: Geo. C. Shreve & Co.,
San Francisco, U.S.A.
Inscribed, 1889
Gold, enamel
$3^3/8$ x $1^3/4$
Inscribed: "Second Prize"
"Light Weight" "1889"
On reverse: "San Francisco Cal."
"July 18, & 19, '89"
"L. Greenebaum"
Gift of Freida Sidorsky
80.74

128 Photograph, Young Men's
Hebrew Association, 1902
$11^1/2$ x $13^1/2$
Jewish Community Center,
San Francisco Collection
Western Jewish History Center

129* Poster, Jewish Folk
Chorus of Petaluma, 1966
17 x 12
Jewish Folk Chorus Collection
Western Jewish History Center

130 Photograph, Camp for
Living Judaism Folk Dancers,
1954
Milton Mann Studios
8 x 10
Florence Freehof Collection
Western Jewish History Center

131 Oakland Police
Department shield, 1955
2 x $2^1/2$ x $1/2$
Rabbi William Stern Collection
Western Jewish History Center

132 Firefighter's helmet
$14^1/2$ x 12 x $9^1/2$
Sherith Israel Collection
Western Jewish History Center

New Interpretations and New Traditions

133 Resolution on wearing
of hats at services, 1910
11 x $8^1/2$
Sherith Israel Collection
Western Jewish History Center

134 Letter regarding use
of microwave ovens in
Kosher kitchens, 1987
$10^1/2$ x $7^1/4$
Florence Freehof Collection
Western Jewish History Center

135 *Jews and Buddhism: Belief
Amended, Faith Revealed* (excerpt)
Bill Chayes and Isaac Solotaroff,
1999
Beta Sp video

136* Jewish Women's
Conference and Fair flyer, 1978
11 x $8^1/2$
Ephemera collection
Western Jewish History Center

137 Womens' skullcaps (*kippot*)
Artist: Edna Sandler
(b. 1963 Capetown, South Africa)
Toronto, Canada, 2000
Dupioni silk, paint;
and rayon thread
Museum purchase
Recent acquisition

161 Photograph of Seder at Emanu-El Residence Club, 1916

This photograph shows women residents and their guests celebrating Passover together at the home of the Emanu-El Sisterhood for Personal Service on Steiner Street and Golden Gate Avenue in San Francisco. The organization provided room and board for single, working Jewish women living apart from their families.

SPRING

HEBREW MONTHS: *Adar II (in a leap year) Nisan, Iyyar, Sivan, Tammuz*
GREGORIAN MONTHS: *March, April, May, June*

Spring is a season of new beginnings: new growth for the land and, metaphorically, for human life as well. Jewish birthing rituals connect babies to their ancestral heritage through *brit milah* (ritual circumcision) for eight-day-old boys and, according to some customs, through baby naming ceremonies for boys and girls.

Passover, celebrated for eight days, commemorates the Exodus, considered the birth of the Jewish people when they were liberated from slavery in Egypt. During Passover each person is obligated to regard him or herself as having personally participated in the Exodus — as if bringing the past into the present. Participants recall in a tangible way the experience of slavery and the drama of redemption through preparation rituals, participating in a *seder* (the ritual meal) during which the *Haggadah* (book of commentaries, prayers, songs, and rituals for the *seder*) is recited, and eating symbolic foods.

Shavuot (Weeks) is celebrated exactly seven weeks after Passover. The ritual counting of the days between the two holidays is called *Omer* (the Hebrew word for a sheaf or measure of grain), recalling an ancient harvest ritual. *Shavuot* commemorates the momentous occasion when, at Mount Sinai, the Ten Commandments were transmitted from God to Moses and Moses to the people to be the foundation of Judaism throughout the ages.

The concept of transmitting the heritage from generation to generation is common to most Jewish observances; it is, however, particularly evident during *bar mitzvah* for boys and *bat mitzvah* for girls. Although customs vary, according to Jewish law this usually occurs at around age twelve for girls and thirteen for boys, when they are considered mature enough to accept responsibility for fulfilling the commandments. They customarily celebrate the occasion by reading a portion of the biblical text at synagogue services; great festivity often follows. ■

144 Outdoor photographs

145 *Boy Awakening,* 1888
Toby Rosenthal

171 *Haggadah* for Passover, 1966
Ben Shahn

left:
162 *Passover Still Life,* 1979
Linda Plotkin

right:
191 *Synagogue Clock,* 1998
Donald Sultan

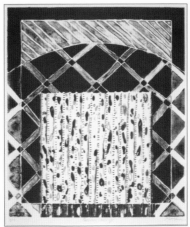

181 Wollega, Ethiopia, 1970s
Marriam Cramer Ring

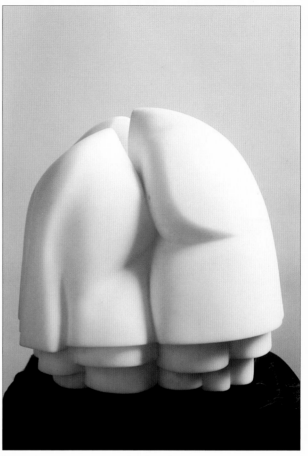

180 *Readers,* 1969
Elbert Weinberg

155 Circumcision bowl, 17th century

left:

172 *Omer* calendar,
ca. 18th century

top right:

183 Plaque commemorating
the *bar mitzvah* of William
Sternsher, 1909

bottom right:

192 *Bar Mitzvah*, 1991,
Ira Nowinski

176 Torah ark from the
Paradesi synagogue,
ca. 1870–1910

SPRING

Nature

138 Israel Art Calendar 1952/53
Corner of Mount Carmel,
March 1953
Publisher: Lion the Printer
Printer: Haaretz
Tel Aviv, Israel, 1952
Paper, printed; and base metal
Judaica collection

139 Untitled (women planting),
mid-1920s
Maurycy Minkowski
(1881–1930 Poland)
Oil on wood
29 1/2 x 24
Gift of the Greta R. Windmiller
Trust
98.2.2

140 Photograph
Bissinger and Company Records
8 x 10
Western Jewish History Center

141 Plaque, showing plowing
of field with oxen
Bezalel School of Arts and Crafts
Jerusalem, Palestine,
ca. 1906–1929
Silver; base metal; mounted
on wood
height 7 3/5
Gift of Mrs. Mary Schussheim
85.35.36

142 Plaque, showing Jewish
pioneer holding agricultural tool
Bezalel School of Arts and Crafts
Jerusalem, Palestine,
ca. 1906–1929
Silver; base metal; mounted
on wood
height 6 1/2
Gift of Mrs. Mary Schussheim
85.35.37

143 Soil samples
Collected by Walter C.
Lowdermilk, March–July 1939
From: Ctesiphon and Babylon,
Mesopotamia; Lebanon;
Jericho, Palestine; Jerash,
Trans-Jordan; and Damascus,
Syria
Jars: Glass; and base metal
5 5/8; diameter: 3 1/4
Jar contents: Soil
Judah L. Magnes Museum
collection

144* Outdoor photographs (2)
7 x 4 1/2 and 4 x 4 3/4
Sophie Gerstle Lilienthal
Collection
Western Jewish History Center

Youth

145* *Boy Awakening,* 1888
Toby Rosenthal
(France 1848–1917 Germany)
Oil on canvas
22 x 21 1/4
Gift of the Lilienthal family
75.30

146 Palestine, ca. 1920–1930
United Jewish Appeal
Gelatin silver print
8 x 10
Museum purchase
80.77.3.20

147 Peter, ca. 1906
Emil Stoger, Austria
Gelatin silver print
4 1/8 x 2 5/8
Gift of Geri Senigaglia
93.39.199

148 U.S.A., 1944
Photographer unknown
Gelatin silver print
2 1/4 x 3
Gift of Geri Senigaglia
93.39.149

149 Guido, ca. 1907
Emil Stoger, Austria
Gelatin silver print
4 1/8 x 2 5/8
Gift of Geri Senigaglia
93.39.192

150 Poland, 1910
Photographer unknown
Gelatin silver print
5 x 3
Gift of Raymond Abrams
93.5.28

Birth: Connecting to Ancestors

151 Amulet for mother
and newborn child
India, 20th century
Watercolor, ink, and graphite
on paper
20 7/8 x 13 1/4
Museum purchase
Rabbi Kimmel Collection
A-28

152 Amulets for newborn
boy and girl
Germany, 18th century
Paper, printed
boy: 9 x 7 5/8
girl: 7 1/8 x 8 5/8
Judaica collection
2000.0.2 & 3

153 Torah binder (*wimpel*)
Gernsheim, Germany, 1857
Linen, painted; embroidered
with cotton thread
7 7/8 x 115 3/4
Museum purchase
Strauss Collection
67.1.21.4

154 Palestinian Jewish seal
7th century B.C.E.
Lapis lazuli
11/32 x 7/16 x 1/8
Inscribed: "Belonging to
Immadiahu daughter of
Shevaniahu"
Museum purchase
Strauss Collection
67.1.12.1

155* Circumcision bowl
Italy, 17th century
Silver, gilded
1; diameter: 4
Gift of Margaret Kaplan in
memory of the Kay and
Landauer families
2000.7.1

156 *La Circoncision*
(from the portfolio
Scènes de la Vie Juive),1884
Bernard Picart (1673–1733 France)
Heliogravure (posthumous
edition)
19⁷/8 x 12⁷/8
Gift of Rabbi Irving F. Reichert
76.223

157 Circumcision,
Brooklyn, 1974
Neil Folberg (b. 1950 U.S.A.)
Gelatin silver print
14 x 16
Gift of David Garfinkle
97.52.19

Passover: Bringing the Past into the Present

158 Baking *matzah*,
Mea Shearim, 1930s
J. Y. Eisenstark, Israel
Gelatin silver print
8 x 10
Museum purchase
80.77.20

159 Poland, ca. 1914
Joint Distribution Committee
Gelatin silver print
5 x 7¹/4
Museum purchase
78.20.18

160 Passover Service
Nancy, France, 1945
Sgt. Gus Lempeotis
Gelatin silver print
6¹/2 x 9
Gift of Abraham Haselkorn

161* Photograph of *Seder* at
Emanu-El Residence Club, 1916
8 x 9¹/2
Emanu-El Residence Club
Collection
Western Jewish History Center

162* *Passover Still Life,* 1979
Linda Plotkin (b.1938 U.S.A.)
Color lithograph
20¹/8 x 17¹/8
Gift of the artist
93.20.2

163 *Seder,* San Francisco, 1991
Ira Nowinski
(b. 1942 U.S.A.)
Gelatin silver print
16 x 20
Museum purchase
92.29.5

164 *Seder* plate
Bezalel School of Arts and Crafts
Jerusalem, Palestine,
ca. 1906–1929
Silver
³/8; diameter: 13
Gift of Mrs. Mary Schussheim
82.20.1

165 *Seder* plate
Manufacturer: Ridgways
Staffordshire, England, ca. 1850
China, glazed; with transfer print
1; diameter: 10
Gift of Connee Stuchinsky
86.25

166 Decorative wall panel
France, 1724/25
Silk damask, embroidered with
metal and silk thread; and braid
24 x 15¹/4
Museum purchase
Strauss Collection
67.1.14.10

167 *Matzah* cover
Russia, 19th century
Cotton, velvet, embroidered
with metal thread; and metal
disks; diameter: 16¹/2
Inscribed in Hebrew:
"Remembrance of the Exodus from
Egypt"; abbreviations for the
names of the ritual foods that are
placed on the *seder* plate
Gift of Mr. and Mrs. B. Sobler
75.183.156

168 Elijah cup
Manufacturer: Steuben
U.S.A., 1979
Glass; 9; diameter: 4⁷/8
Gift of Dr. and Mrs. Phillip Bader
82.7

169 Plaque
Design: Arnold Zadikow
(1884–1943)
Reproduced: U.S.A., 1984
Copper alloy, mounted on wood
8¹/4 x 8⁷/8 x ⁹/16
Gift of Anita and Victor Keyak

170 Service for the Two First
Nights of the Passover
New York: S. H. Jackson, 1837
In Hebrew and English
6¹/4 x 4¹/5
Haggadah Collection
Blumenthal Rare Book
and Manuscript Library

171* *Haggadah* for Passover
Illustrator: Ben Shahn
(1898–1969 Lithuania)
Paris: Trianon Press, 1966
In Hebrew and English
15³/4 x 12
Copy no. 214 of 218 copies
Museum purchase through
donation by Jay Espovich in
memory of Norman Espovich
Haggadah Collection
Blumenthal Rare Book
and Manuscript Library

172* *Omer* calendar
Germany, probably 18th century
Gouache on paper
23¹/2 x 17¹/2
Museum purchase
Strauss Collection
67.1.7.16

Transmitting Judaism from Generation to Generation

Shavuot: The Feast of Weeks

173 New Year's card
Men and boys holding Torah
scrolls in a synagogue
Printed in Germany, late 19th–
early 20th century
Paper, printed
11¹/2 x 8⁵/8 x 4³/8
Museum purchase
96.40.3

174 *Scholar and Student,* 1921
Alfred Lakos
(1870–1961 Hungary)
Oil on board; 14⁵/8 x 11³/8
Gift of Sandra and Michael
Lawrence
95.9

175 *Shavuot,* 1984
Theo Tobiasse
(b.1927 Israel)
Color lithograph with
carborundum etching
and embossing, 82/125
22 1/2 x 30 1/4
Gift of
Mr. and Mrs. Kenneth Nahan
90.17

176* Torah ark from
the Paradesi synagogue,
Ernakulam, India,
ca. 1870–1910
Wood, paint, gilding
140 x 91 x 12
Inscribed in Hebrew:
"Crown of the Torah"
Gift of the Jewish community of
Ernakulam, India
67.0.3

177 Torah mantle (*me'il*)
Morocco, late 19th–early
20th century
Silk velvet, embroidered with
metal thread, cardboard forms;
metal thread braid with cell
fringe; linen and silk lining
34 1/2 x 11 x 10 1/2
Gift of the Bengualid family
75.183.97

178 Prayer shawl (*tallit*)
Eastern Europe, 19th century
Wool, dyed; gold and silver
metallic thread; cotton
Gift of Robert E. Levinson
76.290

179 Fringed garment
(*tallit katan*)
Germany, 19th century
Cotton, crocheted; silk,
embroidered
26 x 8 3/4 (including fringe)
75.183.280

180* *Readers,* 1969
Elbert Weinberg
(1928–1991 U.S.A.)
Marble
15 3/8 x 15 x 13 3/8
Gift of the Elbert Weinberg Trust
98.11.1

181* Wollega, Ethiopia, 1970s
Marriam Cramer Ring, U.S.A.
Gelatin silver print
14 x 11
Gift of the artist
93.9.20

182 Figurine
Wollega, Ethiopia, 1987
Clay
9 x 3 3/4 x 4 3/8
Gift of Marriam Cramer Ring
96.31.1

Bar and Bat Mitzvah: Coming-of-Age

183* Plaque commemorating
the *bar mitzvah* of William
Sternsher
U.S.A., 1909
Ink and metallic paint on paper
photograph
16 1/8 x 13 1/8
Gift of Patricia Stern Green
87.24

184 Wood plaque, 1960
3 x 8 1/2 x 1 1/2
Rabbi William Stern Collection
Western Jewish History Center

185 *Bat mitzvah* invitation
for Alexis Ruben
Poway, California, U.S.A., 1989
Paper, printed
8 1/4 x 8 1/4
Rabbi Mark Hurvitz Collection

186 *Bar mitzvah* invitation
for Chad Travis
Poway, California, U.S.A., 1999
Paper, printed
8 1/2 x 7
Rabbi Mark Hurvitz Collection

187 *B'nai mitzvah* invitation
for Benjamin Nhu and Julie
Megan Seifer
Montclair, New Jersey, 1987
Paper, printed
7 1/4 x 7 1/4
Rabbi Mark Hurvitz Collection

188 *Bat mitzvah* invitation
for Marlene Inselberg
Livingston, New Jersey, 1982
Paper, printed
5 1/4 x 7 1/4
Rabbi Mark Hurvitz Collection

189 *Bar mitzvah* cake
ornament
U.S.A., early 20th century
Bisque, painted; plaster;
and fabric
10 1/2 x 5 1/2 x 3 3/4
Inscribed in Hebrew:
"M[azal] T[ov]"
Gift of Dr. Philip Feiger
99.47.3

190 Disposable camera box
Bat mitzvah memories
Filmart, Brooklyn, New York
1999
Paper, printed
2 1/2 x 5 x 1 5/8
Rabbi Mark Hurvitz Collection

191* *Synagogue Clock*
(from the portfolio *Bar Mitzvah*),
1998
Donald Sultan
(b. 1951 U.S.A.)
Text by David Mamet
Screenprint with 22-karat
gold leaf, 98/395
17 x 22 1/2
Museum purchase with
funds provided by the
Helzel Family Foundation
99.30

192* *Bar Mitzvah,*
San Francisco, 1991
Ira Nowinski
(b. 1942 U.S.A.)
Gelatin silver print
20 x 16
Museum purchase
92.29.18

193 *Eugene* (excerpt), 1994
Rebecca Michelman
(b. U.S.A.)
Beta Sp video

194 Diploma of Judah Leon
Magnes, 1894
22 3/4 x 16 1/2
Judah L. Magnes Collection
Western Jewish History Center

196 *Small Garden*, ca. 1911, Max Liebermann

In this painting, Liebermann captures the fleeting colors and warmth of a particular time of day. This garden is in Wannsee, outside Berlin, at his summer home, which he frequently painted. With works such as this, Liebermann, one of the foremost impressionists in Germany, influenced a whole generation of German artists.

SUMMER

HEBREW MONTHS: *Sivan, Tammuz, Av, Elul*
GREGORIAN MONTHS: *June, July, August, September*

Summer is a time for blossoming and maturing — in nature and in individuals. Trees and flowers are in full bloom and it is a time to enjoy the out-of-doors. As people reach adulthood, some choose to make a lifetime commitment to another person by formalizing their union in a ceremony. Most Jewish marriage rituals and objects symbolize this commitment: the *chuppah* (canopy), under which the couple stands during the ceremony, symbolizing unity, the home, and God's presence in the relationship; the ring, a universal symbol of wedlock, its shape denoting continuity and eternity; and the official Jewish marriage document (*ketubbah*) a signed contract between the couple, reconfirming their commitment and obligations to each other.

On the summer calendar is *Tishah be-Av* (the 9th of Av), a traditional day of mourning and fasting, a day of remembering periods of adversity throughout Jewish history. On *Tishah be-Av* we are reminded that prejudice, racism, and oppression occur in every era and in every generation, and are suffered by other cultures. It is important not only to remember those who have suffered in the past but also to heed the lessons of history by responding as communities to overcome adversity in the present. ■

213 *Jewish Wedding,* late 19th century
A. Trankowsky

220 Letter from
Sir Moses Montefiore to
the Jewish Community
of Frankfurt, May 3, 1841

202 *Seder ha-Tefilot
mi-kol ha-Shanah,*
1790/91

204 Wedding dress and
bridal accessories, 1907

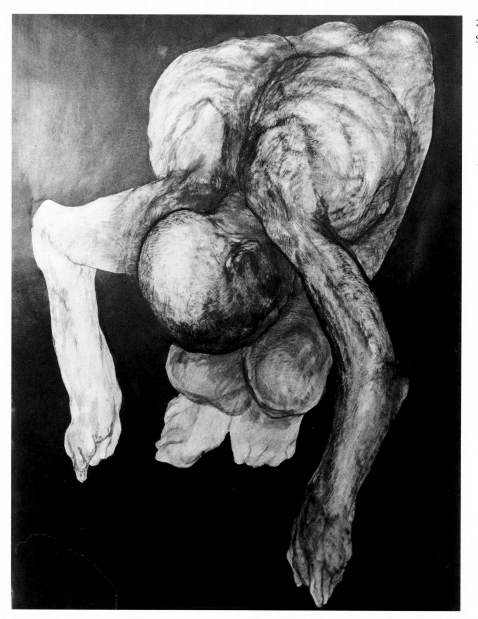

233 *Falling Man,* 1967
Selma Waldman

228 American
Nazi Party rally, 1966

Saul Miller

237 A young Jewish
widow signs a voucher
for cash relief, 1923

236 Jewish Community
leaders negotiating with
Bosnian Serbs, 1994

Edward Serotta

235 *Birth of Jewish Resistance,* 1905, Lazar Krestin

SUMMER

Nature

195 Israel Picture Calendar
1961/2
"FORE! At Caesarea, where
the Romans raced their chariots,
Israelis and visitors play a
relaxing game of golf"
Publisher: Lion the Printer
Printer: Japhet and Lewin
Epstein Presses
Tel Aviv, Israel, July 1962
Paper, printed; and base metal
Judaica collection

196* *Small Garden,* ca. 1911
Max Liebermann
(1847–1935 Germany)
Oil on board
17 1/2 x 21 1/2
Gift of Dr. and Mrs.
Ernst Windesheim
80.36

197 Landscape, ca. 1900
Lesser Ury
(1861–1931 Germany)
Pastel on cardboard
13 3/4 x 19 3/4
Gift of Julius Glaser
92.19

198 Senigaglia family album
Photographers unknown
Mixed media, gelatin silver prints
9 x 12
Gift of Geri Senigaglia
93.39.377–93.39.556

Adulthood: Lifetime Commitment

199 *Lovers,* 1936
Peter Krasnow
(Russia 1887–1979 U.S.A.)
Ebony
20 x 12 x 15
Gift of Peter Krasnow estate
80.16.11

200 *Betrothal,* 1928
Peter Krasnow
(Russia 1887–1979 U.S.A.)
Lithograph
19 1/2 x 15 1/2
Gift of the artist
77.142

201 Mella and Alberto
Senigaglia, ca. 1950s
Photo booth
Gelatin silver print
6 x 1 1/2
Gift of Alberto and
Mella Senigaglia
93.93.306

202* *Seder ha-Tefilot mi-kol
ha-Shanah* (Daily Prayer Book
for the Whole Year)
Amsterdam, The Netherlands:
Yohanan Levi Rofe and his
brother-in-law Baruch, 1790/91
In Hebrew
7 9/10 x 6 1/4
Cover: Silver, repoussé,
by J. L. Kawerstein,
Konigsberg, Germany
Rare Book Collection
Blumenthal Rare Book
and Manuscript Library

203 Wedding invitation
and wedding program for
Gregg Drinkwater and
David Shneer
Berkeley, California,
U.S.A., 1996
Paper, printed
Gift of Gregg Drinkwater
and David Shneer
Recent acquisition

204* Wedding dress and
bridal accessories
San Francisco, U.S.A., 1907
Blouse: silk; cotton lace;
and boning
Skirt: silk
Cummerbund: silk
Shoes: leather;
and plastic button
Gloves: leather; and enameled
metal snap
Stockings: cotton; and silk
thread embroidery
From the wedding of Bertha
Neumann to Herbert Hoppe,
June 9, 1907
Gift of Virginia Hoppe
Ladensohn
91.22.1-6

205 Wedding of Russian Jewish
emigres, San Francisco, 1991
Ira Nowinsky (b. 1942 U.S.A.)
Gelatin silver print
20 x 16
Museum purchase
92.29.22

206 Wedding ring
Italy, 16th–17th century
Gold filigree and enamel; 3/4
Museum purchase
Strauss Collection
67.1.16.1

207 Wedding ring
Italy, 17th–18th century
Gold
diameter: 1
Gift of Peter J. Bickel
83.66

208 Marriage contract
(*ketubbah*)
Lugo, Italy; dated Wednesday,
15 Adar II 5546
(March 15, 1746)
Gouache, metallic paint,
and graphite on vellum
31 x 24 3/4 x 1 1/2
Bride: Bracha, daughter of
the late Raphael Pesach
Groom: Mordecai, son of the
late Joseph Malvecchio
Museum purchase
Strauss Collection
67.1.6.3

209 Covenant of Friendship
(*Berit Re'ut*) contract for
Pauline Abreu Moreno and
Debra Ann Lobel
Maker: Naomi Teplow
Oakland, California, 1997
Watercolor, ink, and graphite
on paper
19 1/8 x 17 1/8
On loan from Pauline Moreno
and Debra Lobel

210 *June Bride* (excerpt)
Sara Felder
(b. 1959 U.S.A.)
Beta Sp video

211 Wedding Party of Abraham
Lazar and Ida Cherkas
Philadelphia
December 10, 1903
Photographer unknown
Gelatin silver print
8 x 10
Museum purchase
97.0.0274

212 India, ca. 1900–1910
K. Barjorji (dates unknown)
Gelatin silver print
6½ x 4½
Museum purchase
97.0.0196

213* *Jewish Wedding,*
late 19th century
A. Trankowsky
Oil on canvas
33 x 47
Purchased with funds donated
by Dr. and Mrs. Fritz Schmerl
in memory of
Professor Hamilton Wolf
75.19

214 *Portrait of Mr. and Mrs.
Hermann,* mid-19th century
Artist unknown
Oil on board
16 x 11¾
Gift of Edith I. Friedman
88.37.1

215 Letter of divorce (*get*)
New York, New York, 1854
Ink on paper
12¾ x 8¾
Husband: Chaim ben Moshe
Wife: Dayla bat Meir
Recent acquisition

216 Letter of divorce (*get*)
Plotzk, Poland, 1649
Ink on vellum
14 x 10¼
Husband: Moshe ben Yaakov Yosef
Wife: Mindel bat Yaakov
Museum purchase
Strauss collection
67.1.15.34

*Tishah be-Av:
A Day of Mourning
and Remembering*

217 *Jüdisches Ceremoniel*
(Jewish Ceremonies); plate no. 7
Author: Paul Christian Kirchner
Illustrator: Johann George Pushner
Nuremberg, Germany:
Peter Conrad Monath, 1726
In German
8¼ x 6¾
Gift of Mr. and Mrs. Harold
Edelstein in memory of
Frederick Kahn, Jr.
Rare Book Collection
Blumenthal Rare Book
and Manuscript Library

218 *Megillath Echa*
(Book of Lamentations)
Berlin: Vorstand der Jüdischen
Gemeinde zu Berlin, 1938
Published for the 9th of Av,
based on the Ludwig Phillipson
translation of the Bible:
Leipzig, Germany, 1859
In German
4½ x 3 1/10
Miniature Book Collection
Blumenthal Rare Book
and Manuscript Library

219 *Tishah be-Av* prayers
found in Cairo, Egypt,
14th century
Ink on vellum
12⅞ x 9¾
Blumenthal Rare Book
and Manuscript Library

*Oppression Occurs in
Every Generation*

220* Letter to the Jewish
community of Frankfurt,
regarding The Damascus Affair
Author: Sir Moses Montefiore
London, England
dated May 3, 1841
Ink on parchment; silk;
and wax seal
26⅜ x 21¾
Museum purchase
Strauss Collection
67.1.17.1

221 Postcard
Portrait of Sir Moses Montefiore
Germany, late 19th century–
early 20th century
Clay-coated paper, printed
6½ x 4¼
Museum purchase
Strauss Collection
67.1.10.25

222 *Lynching,* 1939
Boris Gorelick
(Russia 1909–1984 U.S.A.)
Lithograph
13 x 15⅞
Gift of Leon and
Molly C. Gorelick
97.31

223 Folk art figure
Germany, 19th century
Wood, painted
Inscribed in German
"Father Abraham"
Museum purchase
Strauss Collection
67.1.22.1

224 Periodical illustration
from Puck, vol. 8, no. 196
"The Chosen People. I have
thriven on this sort of thing
for Eighteen Centuries—
Go on, gentlemen,
Persecution helps de Pizness"
Illustrator: J. Keppler
Lithographers: Meyer,
Merkel & Ottman
New York, New York:
Puck Publishing Company,
December 8, 1880
Paper, printed
13⅜ x 19⅞
Gift of Harold M. and
Sandra A. Sofris
82.70.1

225 Illustration from
*The Book of Chronicles from the
Beginning of the World*
(the *Nuremberg Chronicle*)
The "martyrdom" of Simon
of Trent
Hartmann Schedel (1440–1514)
Germany, ca. 1493–1500
Paper, printed
17⅜ x 11⅜
Gift of Seymour Fromer
76.80

226 *Der Sturmer*
"Jüdischer Mordplan gegen die nichtjudische Menschheit aufgedeckt"
(The Jewish plan to murder non-Jewish people is exposed)
Editor: Julius Streicher
(1885–1946)
Nuremberg, Germany: Verlag
"Der sturmer," May 1934
Published weekly, 1923–1945
In German
16¹/₂ x 12¹/₂
Periodical collection
Blumenthal Rare Book
and Manuscript Library

227 Print
"Hepp! Hepp!"
Frankfurt, Germany, ca. 1819
Paper, printed; and gouache
8 x 11
Museum purchase
Strauss Collection
67.1.18.11

228* American Nazi Party rally,
San Francisco, October 23,
1966 (2)
Saul Miller (1917–1989)
Gelatin silver prints
8 x 10
Gift of Richard Miller
in memory of Saul Miller

229 Bernard Zakheim
opposing the Nazi Party rally at
the San Francisco Civic Center,
October 23, 1966
Saul Miller (1917–1989)
Gelatin silver print
10 x 8
Gift of Richard Miller
in memory of Saul Miller

230 Israeli flag, burned
Recovered from Temple
Beth Shalom, Carmichael,
California, 1999
Synthetic and natural fibers
6¹/₂ x 9⁷/₈
Gift of Congregation
Beth Shalom
Recent acquisition

231 Sanctuary chair, burned
Recovered from Temple Beth
Shalom, Carmichael,
California, 1999
Frame: metal; upholstery:
synthetic foam and fabric
35 x 21¹/₂ x 22³/₄
Gift of Congregation
Beth Shalom
Recent acquisition

232 *Shirim* (Poems)
Author: Hayyim Nahman Bialik
(1873–1934)
Illustrator: Joseph Budko
(1888–1941)
Berlin: *Hovevey ha-Shirah
ha-Ivrit* (The Hebrew Poetry
Lovers), 1923
In Hebrew
10 x 8
Illustrated Book Collection
Blumenthal Rare Book
and Manuscript Library

*Remember Those Who
Suffered*

233* *Falling Man*, 1967
Selma Waldman (b. 1938 U.S.A.)
Pastel, watercolor, ink,
and charcoal
58⁷/₈ x 34¹/₂
Gift of Mr. and Mrs. Albert
Schultz
87.55

234 Detail of the Holocaust
Memorial, San Francisco,
by the artist George Segal, 1992
Ira Nowinski (b. 1942 U.S.A.)
Gelatin silver print
13 x 9
Museum purchase
94.4.2.4

*We Heed the Lessons of
History by Responding to
Overcome Adversity*

235* *Birth of Jewish
Resistance*, 1905
Lazar Krestin
(1868–1938 Lithuania)
Oil on canvas
56 x 76
Gift of Copeland Road
Associates
84.55

236* Jewish community
leaders negotiating with
Bosnian Serbs in the barracks
in Lukovice, 1994
Edward Serotta (b. 1949 U.S.A.)
Gelatin silver print
12 x 17¹/₂
Gift of the artist

237* A young Jewish widow
signs a voucher for cash relief,
Riga, Latvia, 1923
Joint Distribution Committee
Gelatin silver print
5 x 7¹/₄
Museum purchase
78.20.6

238 Children waiting to be
examined at the JDC-supported
Jewish Hospital Dispensary,
Kremenchug,
former USSR, 1921
Joint Distribution Committee
Gelatin silver print
5 x 7¹/₄
Museum purchase
78.20.7

239 *Not in Our Town*,
1995 (excerpt)
California Working Group,
Oakland, California
Beta Sp video

240 T-shirt
Assembled in Dominican
Republic of U.S.A. components;
obtained in Sacramento,
California, 1999
Cotton, polyester, and ink
Gift of Congregation Beth Israel
Recent acquisition

126 *Challah* cover, inscription date 1851/52

Inscribed in Hebrew: "Gedaliah son of Simon Ha-Levy known as Ullman"; "Mark that the Lord has given you the Sabbath" (Exodus 16:29); in Latin: "One out of many."